Reaching for the Sun

How Plants Work

From their ability to use energy from sunlight to make their own food, to combating attacks from diseases and predators, plants have evolved an amazing range of life-sustaining strategies.

Written with the non-specialist in mind, John King's lively natural history explains how plants function, from how they gain energy and nutrition to how they grow, develop, and ultimately die. New to this edition is a section devoted to plants and the environment, exploring how problems created by human activities, such as global warming, pollution of land, water and air, and increasing ocean acidity, are impacting on the lives of plants.

King's narrative provides a simple, highly readable introduction, with boxes in each chapter offering additional or more advanced material for readers seeking more detail. He concludes that, despite the challenges posed by growing environmental perils, plants will continue to dominate our planet.

JOHN KING is Professor Emeritus of Biology at the University of Saskatchewan. He is a Past-President of the Canadian Society of Plant Physiologists and in 2001 he was awarded their highest honor, the Gold Medal, "in recognition of outstanding contributions to plant physiology in Canada." The first edition of *Reaching for the Sun* (Cambridge, 1997) was nominated for the Rhône-Poulenc (now Aventis) Prize for Science (General Category).

Reaching for the Sun

How Plants Work

Second Edition

JOHN KING
University of Saskatchewan, Canada

CAMBRIDGE
UNIVERSITY PRESS

CAMBRIDGE UNIVERSITY PRESS
Cambridge, New York, Melbourne, Madrid, Cape Town, Singapore,
São Paulo, Delhi, Dubai, Tokyo, Mexico City

Cambridge University Press
The Edinburgh Building, Cambridge CB2 8RU, UK

Published in the United States of America by
Cambridge University Press, New York

www.cambridge.org
Information on this title: www.cambridge.org/9780521518048

First edition © Cambridge University Press 1997

Second edition © John King 2011

First published 1997

Second edition 2011

Printed in the United Kingdom at the University Press, Cambridge

A catalog record for this publication is available from the British Library

Library of Congress Cataloging in Publication data
King, John, 1938-
   Reaching for the sun : how plants work / John King. – 2nd ed.
      p.   cm.
   Includes bibliographical references and index.
   ISBN 978-0-521-51804-8 (hardback) – ISBN 978-0-521-73668-8 (pb)
   1. Plants.   2. Botany.   I. Title.
   QK50.K46 2011
   571.2–dc22   2010039934

ISBN 978-0-521-51804-8 Hardback
ISBN 978-0-521-73668-8 Paperback

Contents

Preface to the Second Edition

The rationale for the 1997 edition was that, the more people that know about the lifestyle of plants, the more likely it is that they will appreciate what has to be done to preserve this component of the biosphere upon which our survival depends. A secondary aim was the hope that people would discover that knowing more about how plants work is not just useful but also fun.

This second edition is addressed primarily to an audience of students of plant sciences but also to keen gardeners, naturalists, or anyone with questions about why the Earth is green, how plants defend themselves against diseases and predators, how they combat the stresses of constant exposure to the environment, and how climate change is affecting plants.

The general alarm voiced in the 1997 edition about the deleterious effects humans are having on our environment through global warming, deforestation, overgrazing by domestic animals, overcropping of arable lands, and pollution of land, water, and the atmosphere, is now, more than 10 years on, more clearly focused. Thus, in addition to making the book's original contents consistent with the most recent knowledge, material has been added to educate students of all ages about how human activities are impacting the lives of plants – how plants are affected at the physiological level by changes to our environment, such as through increasing concentrations of greenhouse gases, higher temperatures, longer growing seasons, increasing ocean acidity, changing water economy of the Earth, and pollutants such as ozone. Knowledge of influences such as these on the lives of plants is an essential component in the education in the twenty-first century of anyone with an interest in the green component of our planet that is so critical to our survival as a species.

Chapters are grouped into five parts, the first four of which are made up of topics carried over from the 1997 edition. The original 17 chapters have been reduced to 14, mainly by merging and culling material. Each chapter has added to it a box containing additional, more advanced, or new information. This extra source was added for the benefit of the more advanced reader or for those seeking more detail.

Part V is entirely new and highlights, in two chapters, the state of the Earth's environment before and after the influence of human activities became increasingly globally significant. This approach is designed to make it easier for the reader to understand how the lives of plants are being affected, positively and negatively, by the accelerated changes to their environments caused by human actions.

Reference material is grouped at the end of each part, is much more extensive than in the 1997 edition, and each citation has beside it a guide as to how it might be useful in amplifying and enriching knowledge provided in the text.

The book ends with a discussion drawing attention to the importance that genetic engineering is likely to have in future in adapting crop plants to climate change and in transforming our world through a new field of human endeavor, synthetic biology.

I am particularly grateful to Marlynn Mierau for his expert help with the illustrations and to Drs. R. B. Horsch, B. R. Neal, and J. W. Sheard, for their critical reading of all or part of the chapters in part V. Their assistance and advice were of great value, but responsibility for the content of all chapters remains mine, alone. I also acknowledge Jacqueline Garget, Katrina Halliday, and Lynette Talbot, the staff members at Cambridge University Press, who, at various times, shepherded the project along with great expertise, patience, and tact.

John King,
Saskatoon, 2010

Preface to the First Edition

The idea for this book arose from a conversation I had one day with two of my neighbors. Neither is a plant specialist but both are keen gardeners. One has in his garden a number of large trees of which not everyone is an unqualified admirer, including the other neighbor. For one thing, the trees shade adjacent gardens from direct sunlight for much of the day, including that of the second neighbor, a somewhat sensitive matter at this northerly latitude where there is a relatively short growing season. In the autumn, immense numbers of leaves find their way into the general neighborhood, often late in the season since some of these particular trees continue shedding leaves even after the first snow. The task of cleaning up frozen, congealed, decaying leaves is not universally appreciated.

Not for the first time, then, the owner of one of the shaded gardens was trying to persuade the tree-loving neighbor to remove his trees which, to the former, were obstacles to productive gardening. As the conversation developed, it became obvious that the aggrieved party thought the main bulk of a tree came from the soil since he made repeated reference to the fact that the offending trees were taking in significant quantities of nutrients through their roots. Of course, plants do absorb many essential minerals and water from soil but we have known for a long time, for more than 200 years in fact, that air, not soil, is the source of the main building block (carbon) from which the bulk of green plants is manufactured.

This experience led me to wonder how many other aspects of the ways in which plants grow and develop are less than well known to non-specialists. Green plants (in particular, those growing on land rather than in water) have a highly specialized lifestyle due to the fact that they are generally fixed in one location throughout their lives. In addition, they contain a unique molecule, chlorophyll, which sets them apart from all other living organisms and gives them the option of manufacturing enormous quantities of their own food through photosynthesis.

Their general dependency on soil as an anchor and for essential mineral nutrients, as well as water, has led to the development of an elaborate root system with many unique functions. Like all living things, plants need water but, unlike many other organisms, they are not able to go searching for it beyond their immediate location. Thus, green plants have evolved a range of devices to obtain, transport, and conserve water as well as ways to combat the effects of excessive wind, drought, cold, heat, and light from which they cannot hide.

Green plants face serious challenges connected to the changing seasons. Flowering, seed production, dormancy, germination, leaf fall, and death, to name some of the more important milestones in the life cycle of plants, all are related to the seasons. Ways to measure time have evolved so that the activities of plants fit into the pattern of seasonal changes in their environment.

Green plants produce an enormous array of elaborate chemicals for only some of which a purpose is known. Plants use them to add color, fragrance, and flavor to their flowers and fruits, to wage war on predators and disease organisms, and to out-compete near neighbors. We make use of many of them ourselves, as cosmetics and pharmaceuticals for example.

These, and other features of green plants, together constitute what biologists would recognize as plant physiology – how plants work. My purpose here is to try to create an interest in and explain in straightforward language to the inquisitive layperson how plants function.

Green plants are all around us. They are the most successful of all living things evidenced by the fact that they are overwhelmingly the most abundant kinds of organisms on earth. We are absolutely dependent on them for food; we cultivate them for our pleasure; and we have used them in a vast number of ways down the centuries to our advantage. There is growing concern that we are following practices that are a serious threat to green plants. For example, we do not know what effect depletion of the ozone layer in our atmosphere will have on plants (or other creatures for that matter); the so-called "greenhouse effect" is also an unknown quantity. We do know that we are destroying vast numbers of plants through practices such as the burning of forests, overgrazing by our domestic animals, and overcropping. The advent of biotechnology and its use in agriculture is causing concern that we may be manipulating natural plant genetic processes in ways of which we are alarmingly ignorant. Overpopulation and the parallel outcome of more intensive agricultural and industrial practices (such as pollution) that go hand in hand with our burgeoning world population are a growing threat to the green plant world. Such problems should concern us all.

Here, I have tried to provide examples of some of the important aspects of how plants work with the rationale that the more people know about the lifestyle of plants, the more likely it is that they will appreciate what has to be done to preserve this component of the biosphere without which we, ourselves, could not survive on our Earth. I also hope that people who read this will discover that knowing more about how plants work is not just useful but also fun.

I would like to acknowledge, with gratitude, the fact that much of the reading and writing for this project were accomplished during a sabbatical leave granted by the University of Saskatchewan during 1994–95. In addition, among those who helped along the way, I wish to make special mention of two: Dr. Timothy Benton, Popular Science Editor at Cambridge University Press, who gave thoughtful guidance at every step with great good humor and tact; and my wife, Myrna, who not only read and commented on each chapter but also provided ideas to help dispel the kinds of mental vapors that, surely, shroud the minds of most writers from time to time.

John King,
Saskatoon, 1997

Part I   Plants and energy

'From dust you came, to dust you shall return,' is one sober, biblical reminder that complex organisms are built from simple chemical elements to which they will revert. From the beginning to the end of their lives, living things wage a battle against natural forces which break down their highly organized structure:

- At the cell level, complex molecules such as proteins and nucleic acids, to name but two of many hundreds, are continually destroyed by hydrolysis
- Valuable molecules are lost to the environment and have to be replaced because cell membranes are leaky
- Our atmosphere is dominated by the highly reactive molecule, oxygen, as a result of which everything on Earth, organic and inorganic, is subject to corrosive oxidation.

Yet, on all sides, we observe organisms using simple materials from their surroundings to maintain, renew, and build complex structures, to achieve which they need a constant supply of energy.

Organisms have evolved two ways of satisfying their absolute need for energy. The most crucial, **photosynthesis**, traps light energy from an outside source, the Sun, to fuel the building of complex organic structures from simple inorganic materials. The other, **respiration**, requires that there be a constant source from which chemical energy can be extracted and used for maintenance and construction.

Photosynthesis and respiration together comprise **bioenergetics**, how living organisms gain the supply of energy they need, which is the subject of Part I.

1 Photosynthesis: the leitmotiv of life

INTRODUCTION

Those who answer gardening questions from the general public will tell you that surprising numbers of people have a basic misconception about plants. The belief that plants build themselves from the soil is widespread even today, more than 300 years after proof showing this not to be so. Why such a belief still exists is puzzling. Consider the common practice of removing lawn clippings. If grass was simply built from soil, a lawn from which kilograms of clippings were removed during the growing season for the past dozen years would resemble a sunken garden, but it does not. Something additional to soil must go into building a plant.

PHOTOSYNTHESIS: THE KEY

We now understand that plants construct themselves from carbon dioxide (CO_2), water, and minerals with the aid of light energy. What plants make by this **photosynthesis** (putting together by light) is an endless supply of carbohydrates: sugars, starch, and cellulose.

Other green organisms can also photosynthesize

Plants are not the only organisms able to photosynthesize. Our oceans, lakes, and rivers are populated by a wide array of green organisms such as those algae which appear as green scum on ponds and lakes; the larger green, brown, and red algae, the seaweeds, found on or near seashores; and other microscopic organisms, the phytoplankton (certain bacteria, diatoms, dinoflagellates, and the smallest algae), which are especially abundant in our oceans.

Green organisms are sources of fuels

How plants and other photosynthetic organisms build themselves from a few inorganic substances is fascinating for practical reasons. Our oil and natural gas resources come from carbohydrates formed millions of years ago, mainly by ancestors of modern day ocean phytoplankton; ancient vegetation (mostly cellulose) is the source of our coal reserves; and plants and phytoplankton are also the primary sources of the foods that fuel the great majority of other living things.

How do they do it?

Attempts to understand how green organisms capture light energy, use it to split water, releasing hydrogen and oxygen, then use the hydrogen to reduce carbon dioxide to form sugars and other carbohydrates go back centuries. Today, given the problem of climate change, there is even a hope that we can find a way to exploit photosynthesis on an industrial scale to split water, giving us a supply of energy from a limitless, endlessly renewable, environmentally friendly source.

Understanding photosynthesis is, then, a matter both of curiosity about how green organisms do something so miraculous with such ease and because of its possible practical value as we phase out the use of fossil fuels such as coal, oil, and natural gas reserves. But first, a word about energy and its sources among organisms.

ENERGY: EVERY LIVING THING REQUIRES IT

Energy drives everything that all living things do. As humans, we understand the need for energy to do physical work. Less obviously, we also need it for our brains to plan, or hope, or dream during sleep. Of the roughly 100 W of energy it takes to drive the entire human body, about one-fifth is used in activities of the brain.

Fuel and heat

The use of any fuel as a source of energy, whether it be in a car, a light bulb, a computer microchip, or a living organism, always leads to the production of heat as a byproduct. On average, about 90% of the energy in any source of fuel ends up as heat as the fuel is used. In warm-blooded animals like ourselves the production of heat is obvious. Less clear is the fact that heat loss occurs in all living organisms, whether warm to the touch or not, plants included. All organisms need fuel to replace this lost energy.

Sources of fuel – Antoine Lavoisier's great insight

One of the earliest explanations of the use by living things of fuel as a source of energy came from the eighteenth-century chemist, **Antoine Lavoisier**, when he wrote:

> Respiration [breathing] is merely a slow combustion [burning] of carbon and hydrogen, which is similar in every respect to that which occurs in a lighted lamp or candle, and, from this point of view, animals that breathe are really combustible bodies which are consumed.

Lavoisier's point was that animals had in them fuels containing carbon and hydrogen that could be slowly "burned" in respiration, releasing energy. This brilliant insight did Lavoisier not the slightest good. He was guillotined during the French Revolution, the judge allegedly dismissing him with the sentiment, "The Republic has no need of savants or of chemists".

Organic compounds

Today, we understand in a way not possible in Lavoisier's time that the thousands of natural chemicals found in living things contain carbon; most of them contain hydrogen as well. They were called **organic compounds** because it was thought at first that molecules containing carbon could be produced *only* by living organisms. We know now this is not so. Millions of organic compounds, many of which are artificial, are quite unlike any of their natural counterparts. They are still called "organic" because they all contain carbon; no other chemical element comes close to matching the diversity of compounds that carbon is capable of forming.

Carbohydrates

Carbohydrates, made up of carbon, hydrogen and oxygen, are among the most important fuels found in living organisms, and are "burned" in respiration to release energy. Those familiar to us are: **sugars** like **glucose** (grape sugar), **sucrose** (table sugar); **starch** (potatoes, corn, cereals, sorghum); and **cellulose**, the strong fibre from which the plant's basic structure is made, but as a fuel is less familiar since we, along with most other animals, cannot digest it. Only ruminant animals, such as cattle, sheep, and deer, and some others, like horses and termites, have adapted to the indirect use of cellulose as a fuel. All have bacteria in their digestive systems which can break down cellulose to a form the microbes can use as fuel. Animals harboring these microorganisms benefit by stealing the surplus fuel not needed by the bacteria while providing an environment in which they can prosper.

Glucose, the lowest common denominator

Before being used in respiration, sucrose, starch, and cellulose are all broken down to glucose. A molecule of sucrose contains one glucose and one fructose unit; starch and cellulose are made up of chains of hundreds of glucose molecules; all can be acted on by enzymes to release glucose.

BACK TO PHOTOSYNTHESIS

Making glucose from CO_2 and water in photosynthesis takes just as much energy as is released in "burning" it in respiration. There's no shortage of these two molecules in the atmosphere, oceans, lakes, the soil, and inside living organisms, but the chance of any of them coming together in the right way to produce even a single molecule of glucose is remote. Such an event is unlikely to have happened in the billions of years CO_2 and water have existed on Earth. The reason for this is that the chemical bonds holding hydrogen and oxygen together in water molecules, and carbon to oxygen in CO_2, are extremely strong and must be broken before these two compounds can interact to form glucose.

SO ... HOW DOES PHOTOSYNTHESIS WORK?

Green plants form glucose from CO_2 and water every daylight hour during the growing season. They then go on to produce a seemingly endless supply of sucrose, starch, and cellulose from the glucose. To appreciate more fully how plants do this so effortlessly, it is useful to know some of the history behind our understanding of photosynthesis.

The history

Lavoisier may have been one of the first to study animal respiration and to understand the need animals have for energy, but long before his day, it was understood that animals must eat to live.

Until some 350 years ago, it was thought that plants obtained their food in a similar way to animals, that is, by "eating" whole matter from the soil. Not only that, plants were also judged to have *perfect* nutrition since, unlike animals, they excreted no waste from what they "ate."

Also, the belief was that all things were made up of "the four elements" – earth, air, fire, and water. Air and fire were thought to have no weight and so it was considered that anything with weight could come only from either earth, water, or both.

Enter Jan Baptista Van Helmont

These early ideas began to change during the seventeenth century. In one of the first recorded scientific experiments, **Jan Baptista Van Helmont**, a Belgian physician, measured the growth of a willow sapling.

- He weighed it at the beginning and end of a 5-year period and found it had gained nearly 75 kg, yet, the soil in the pot in which the plant was grown lost only a few grams

• Van Helmont concluded that, therefore, the increase in mass of his tree could not possibly all have come from the earth in which it was grown since the soil lost only a tiny fraction of the weight gained by the plant.

But then Van Helmont made a mistake. He reasoned that if most of the weight gained by the willow was not from the soil then it must be from the water added to the tree during the 5-year experiment. He was a victim of the state of knowledge of the natural world at the time. As the elements water and earth were thought to be the only sources of bulk, it followed that if earth (soil) was not the source of the gain in weight of the willow, then it had to come from water.

Van Helmont was right in concluding that water contributed to the weight gain by his tree. His mistake was thinking that water was the *main* source of it. But what was particularly startling at the time was that the soil contributed so *little* to the weight gain. This inescapable conclusion changed for ever thinking about where plant food came from.

Farewell to "the four elements": alchemy and Joseph Priestley
There was no possibility of understanding Van Helmont's results until knowledge of chemical elements improved. Progress had to be made in thinking beyond "the four elements." At the forefront of these advances was the late eighteenth-century English clergyman, **Joseph Priestley**.

Combustion, or burning, was a topic that had intrigued alchemists for centuries, especially how one chemical compound could be transmuted into another by heat. The search for ways to transmute baser metals into gold or to produce the "elixir of life" are examples of the fascination alchemists had with the effects of combustion on substances. Modern day industries based on the purification of metals and their fusion together by heat to form alloys are outgrowths of the activities of alchemists, Priestley among them.

Injured and restored air

Priestley was fascinated by the "injuring" of air by combustion and performed some bizarre experiments to demonstrate it; for example:

- if he burned a candle in an air-tight container the flame soon went out;
- if he then put a mouse into the container, the animal died, because the air in the container, he concluded, had been "injured";
- he found, however, that if he put a sprig of fresh mint into an air-tight vessel containing injured air, the air was soon "restored" to a state in which it would "… neither extinguish a candle nor was it at all inconvenient to a mouse, which I had put into it."

In other words, the mouse did not die, was not "inconvenienced," as Priestley so delicately put it, as long as the air in the container was "restored."

Priestley did not see the light …

Priestley's conclusion was that vegetation "restored" air by cleansing and purifying it, removing the "injury." What he did not understand was that these restorative powers of plants depended on light. He used glass vessels in experiments so he could see what was going on inside but never did understand the significance of his choice.

… but Jan Ingen-Housz did

The importance of light in the "restoration" of "injured" air was left to the Dutch physician, **Jan Ingen-Housz**, to discover a few years later. He found air to be "mended" by vegetation only in sunlight and only by the green parts of plants. He also discovered that a plant absorbs the carbon in CO_2,

> throwing out at that time the oxygen alone, and keeping the carbon to itself as nourishment.

Light and air finally linked

Here was the first hint that light and CO_2 were linked in some way, leading to the release of oxygen, and that oxygen was the "restorative" in "injured" air. Here also was a hint that air provided a chemical element, carbon, to the nourishment of a plant. These, and similar, results made it clear that there was more to air than had been imagined. Perhaps it was, after all, possible to gain weight even from something so insubstantial.

We now know that Ingen-Housz was wrong in suggesting that oxygen comes from CO_2. In photosynthesis, it is released during the splitting of water.

Nicholas Theodore de Saussure connects the dots

One further essential observation was needed before the way forward to a full understanding of "synthesis by light" could be discerned.

The Swiss scientist, **Nicholas Theodore de Saussure**, made the final connection in the first decade of the nineteenth century. He showed that:

- in the light a plant released exactly as much oxygen as it absorbed carbon dioxide;
- equally importantly, he showed that the weight a plant gained was greater than could be explained by the amount of carbon taken in as CO_2.

In other words, in addition to CO_2, something else contributed to the solid substance of a plant. This extra contribution, de Saussure showed, had to be from water, since he was able to eliminate all other possibilities.

Van Helmont not so wrong after all

Water is critical to weight gain in a plant, but it is not the nearly exclusive source of the increase that Van Helmont had thought. Taken together, these early studies made it clear that the main bulk of plants could, after all, be produced largely from something as apparently weightless as air, in light only, with the help of water.

The realization of the importance of CO_2 and the need for sunlight for weight gain in plants shifted attention in the study of photosynthesis away from the soil to where it belonged, to the role played by the atmosphere and by light.

LIGHT AS ENERGY SOURCE

The heat of summer and sunburn remind us that there is no lack of energy in sunlight.

- The amount of solar radiation reaching Earth each year is about 4×10^{18} joules (4 exajoules).
- The amount arriving at our planet each hour, some 4×10^{14} joules, is more than enough to satisfy the world's energy needs for an entire year.
- About 60% is reflected directly back into space (recall the images of the brilliant, shining planet Earth as seen from outer space); most of the rest is absorbed by the atmosphere, by clouds, or by oceans and landmasses and promptly re-radiated back into space as heat.
- The amount of sunlight absorbed by green plants and used in photosynthesis is tiny, no more than 1% even of the fraction of solar energy that penetrates to the planet's surface.

ABSORPTION OF LIGHT

Since the late nineteenth century it has been known that, when light is absorbed by metals, electrons are dislodged from them and can be organized into an electric current; solar panels operate on this principle.

Plant pigments

In a plant, light is not absorbed by metals but by **pigments**, molecules whose bright colors signify that they strongly absorb only some of the wavelengths of visible light. The dominant pigment found in plant leaves is chlorophyll, which absorbs the blue and red wavelengths of light; wavelengths which are reflected from the leaf or pass through it give the pigment its characteristic green color. Several other pigments are also found in most plants but cannot be seen until the autumn when chlorophyll disappears from leaves. All of these pigments absorb bluish–violet light so we see them as being yellow, orange, or red, depending on which wavelengths they absorb (see Chapter 11 for a discussion of plant pigments and color, and Chapter 9 where a quite different pigment, phytochrome, is described).

Making sugars, starch, and cellulose

Just as in the case of metals, absorption of light by pigments, principally chlorophyll, results in the following.

- It dislodges electrons, which are then organized within leaf cells into tiny electric currents
- This energy source from light is then used by plants to power a complicated sequence of chemical reactions, with the net effect of splitting water molecules to yield hydrogen and oxygen
- The outcome is the release from the leaf into the atmosphere of oxygen, of hydrogen as a source of reducing power, and the synthesis of an energy source, adenosine triphosphate (ATP; see Box 2, Chapter 2) to fuel the formation, first, of the 3-carbon sugar, phosphoglyceraldehyde, and then the 6-carbon sugar, glucose, from CO_2
- Some of the glucose is used as fuel in respiration to supply energy to a plant for its own life functions just as happens in all living organisms (see Chapter 2)
- However, during most days in a growing season there is light energy enough for leaves to produce more glucose than a plant requires for its immediate needs. Surplus glucose is diverted to produce storage carbohydrates, such as sucrose or starch, and to manufacture cellulose fibres for building plant structures (Box 1).

BOX 1. KEY ADVANCES IN UNDERSTANDING PHOTOSYNTHESIS

The general formula for photosynthesis is often written either schematically as:

$$6CO_2 + 6H_2O = C_6H_{12}O_6 + 6O_2........(1)$$
$$Glucose$$

or empirically as:

$$CO_2 + H_2O = CH_2O + O_2..........(2)$$

Both leave the impression that photosynthesis is a single reaction but this is not so. More detailed evidence about the process began appearing in the late 1920s when Stanford University

BOX 1. (cont.)

microbiologist, **Cornelius Bernardus van Niel**, showed, experimentally, that purple sulfur bacteria had a form of photosynthesis which conformed to the equation:

$$CO_2 + 2H_2S = CH_2O + H_2O + 2S.......(3)$$

where water was not a substrate in the reaction, although some was formed, hydrogen sulfide acted as a hydrogen donor to reduce carbon dioxide, and sulfur was an oxidized byproduct, not oxygen. These results demonstrated clearly that photosynthesis was an oxidation–reduction process. The close resemblance of equations (2) and (3) to one another led van Niel to hypothesize that all forms of photosynthesis could be described by the general equation:

$$CO_2 + 2H_2A = CH_2O + H_2O + 2A.......(4)$$

In most green organisms today, H_2O (H_2A in equation 4) is the preferred hydrogen donor for reduction of CO_2 to carbohydrate and O_2 (A in equation 4) the oxidized product.

In the 1930s, Cambridge University researcher and Nobel laureate, **Robert [Robin] Hill**, demonstrated that isolated leaf chloroplasts gave off O_2 in the light in the absence of CO_2 if supplied with salts containing ferric ion (e.g. ferric oxalate). Hill concluded that the splitting of water in photosynthesis could be induced to proceed in the absence of CO_2 reduction as long as the hydrogen released from the water was removed, e.g.

$$4Fe^{3+} + 2H_2O \xrightarrow[chloroplasts]{light} 4Fe^{2+} + 4H^+ + O_2.......(5)$$

The ferric ion (Fe^{3+}) in equation (5) accepts electrons from hydrogen and is reduced to ferrous ion (Fe^{2+}); reactions of this kind are examples of the so-called **Hill reaction**.

Hill's protocols and results were all-important in opening a way to separating those reactions in the photosynthetic pathway

BOX 1. (cont.)

*dependent on light (the splitting of water to provide the energy and reductive power to the process; the so-called **light reactions**) from those leading to the reduction of CO_2 to carbohydrate (the **dark reactions**), which do not require light. They also led to the eventual development of the ways and means to elucidate the entire process of photosynthesis at the molecular level.*

PHOTOSYNTHETIC EFFICIENCY: THE LEAF

Plants are found everywhere from Arctic tundra to tropical forests to arid deserts. It is not surprising, then, that if we look at a wide array of species we find clear evidence of how green plants have evolved and adapted to make this most critical of all life processes as efficient as possible in a wide range of environments.

Exterior design

Photosynthetic activity in most plants occurs in the leaf, which is beautifully designed for this function. Leaves can vary enormously in size and shape but are most often flat and thin. This gives to a leaf maximum surface area and minimum volume, a design that allows it the greatest area on which to receive maximum light and CO_2 from its surroundings with minimum waste of thickness. Layering of leaves in the canopy of the plant maximizes exposure of the greatest number of leaves to sunlight. Leaves of some plants are also capable of adjusting their position to follow the sun as it moves across the sky, a process called **solar tracking**. All these adaptations aid photosynthesis.

Interior design

The interior of a leaf is also highly specialized for light absorption. The region of the leaf just below its upper surface is made up of **palisade cells**, so named because they are narrow and tall like boards in a palisade fence. They stand in tightly packed, parallel rows one to three layers deep. We may wonder how efficient it is for a plant to develop more than one layer of these cells when we might imagine that the first layer would absorb most of the sunlight. In fact, more light than might be expected goes through the first palisade layer because of what are called **sieve** and **light guide effects**.

The sieve effect

The sieve effect is caused by chlorophyll not being evenly distributed within leaf cells. To the naked eye a leaf looks uniformly green, but if we magnify its interior using a microscope we find that leaf cells contain many tiny green packages called **chloroplasts**; the remaining contents of each cell are colorless. This clustering of pigments leads to not all chlorophyll molecules in each chloroplast being equally exposed to sunlight; some are shaded by others. Not every chlorophyll molecule, therefore, has the same opportunity to absorb light. Also, some light passes between chloroplasts through the clear areas of a cell – it is not intercepted. Thus, some light is not absorbed by the first layer of palisade cells but travels on to the second and third layers: the sieve effect.

The light guide effect

Light guiding is the channeling of some light between palisade cells, deep into the leaf. Because palisade cells are arranged in such tightly packed, orderly layers, the narrow spaces between them act as optical fibers, "light pipes," which, in some smaller seedlings, may direct light all the way to roots. Some of the light entering the upper side of a leaf is directed in this way towards the lower side, the **spongy tissue**, where cells are irregularly shaped and, because of that, have large air spaces between them. Such an arrangement produces many surfaces from which light can bounce and be scattered. We can see something of the effect of this scattering with the naked eye. Look at the lower surfaces of many kinds of leaves and you will find them to be lighter green than the upper surfaces; some of this lighter shade is because of light scattering inside the leaf.

The result

The contrasting arrangement of cells in the palisade and spongy layers – the former allowing some light to pass through and the latter trapping as much light as possible – allows a leaf to intercept and use in photosynthesis the maximum amount of the sunlight entering from above.

CAPTURING CARBON DIOXIDE

Capturing light as efficiently as possible is only one of the challenges facing the photosynthesizing leaf. Equally important is the capture of CO_2 from the air and its transport to where photosynthesis occurs. A major problem is the low concentration of this gas in the atmosphere even as human activities are causing it to increase (see Part V).

Oxygen is at a concentration of about 210 000 parts per million (ppm) in air. Since the dawn of the industrial age, the concentration of CO_2 in the atmosphere has increased from about 280 to 385 ppm, and is rising. Compared to oxygen, there is little CO_2 available to a plant. Thus, anything which improves the efficiency of a leaf to take in CO_2 from the small amount available in the atmosphere favors photosynthesis.

Stomata, carbon dioxide, and C_3 photosynthesis

CO_2 enters a plant through the stomata, the pores found, most often, on the undersurfaces of leaves (see Chapters 3 and 12). Each leaf has thousands of stomata, which remain open all day as long as the plant is well supplied with water. CO_2 diffuses in through the stomata to the cells of the spongy and palisade layers where, in about 97% of plant species, it is used immediately in photosynthesis, in the light. This is referred to as **C_3 photosynthesis** because the first compound formed in the pathway leading to sugars is phosphoglyceric acid (PGA), which contains three carbon atoms.

Adaptation to heat and drought – C_4 photosynthesis

The remaining 3% of plants have a different, two-phase, system of photosynthesis, an adaptation linked to where they originated (see also Chapter 12).

Plants like corn, sorghum, and sugarcane evolved in tropical grasslands, environments with long periods of hot, dry weather. During prolonged periods of heat and drought, the stomata of plants in areas like these are often either completely or partially closed for much of the day to slow down water loss. This severely limits the ability of these plants to take in CO_2, therefore, but they have developed a way of sucking the gas into their leaves efficiently even through partially closed stomata, delivering it, later, in high concentration to where photosynthesis is occurring.

In a first phase, CO_2 is drawn through the stomata into leaf cells where it is incorporated, concentrated, and stored in one or more compounds, all of which contain four carbon atoms. Hence, this type of carbon assimilation is called **C_4 photosynthesis**. In a second phase, CO_2 is released from storage when and where it is needed for photosynthesis in the same pathway, beginning with PGA, as is found in all other plants.

Extreme adaptation to heat and drought

A more extreme example of adaptation to heat and drought is found in cacti and other kinds of fleshy plants which grow in deserts. Here, the shortage of water is often so acute that stomata remain closed all day. These plants minimize water loss by opening stomata only at night when they take in CO_2 and store it in C_4 compounds. Throughout the following day they release the gas from storage, behind closed stomata, as needed for photosynthesis. Pineapple is the best known food plant to have evolved this trick.

SUMMARY

- The realm of living things is distinct from the physical world in organization and how it works
- Yet, the organic world is dependent on the inorganic to maintain its life processes
- Important among these means is the need for carbon and hydrogen in forms that will serve as fuel to burn with oxygen to release energy
- However, in the inorganic world, carbon and hydrogen are locked into two of the most stable, unreactive substances known – carbon dioxide and water
- It is here that plants and other photosynthetic organisms play a crucial role in the drama of life
- Plants can capture light, convert it to chemical energy, then use this energy to reshuffle the atoms in carbon dioxide and water to make carbohydrates, the most basic of all foods; plants are highly efficient carbohydrate factories
- At the same time, plants release oxygen, an essential ingredient for the "burning" of foods to supply the energy needs of all organisms

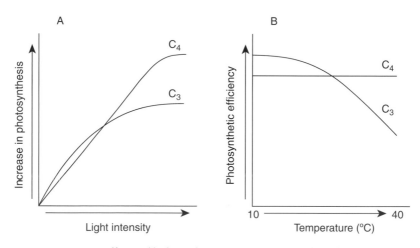

FIGURE 1 Effects of light and temperature on C_3 and C_4 photosynthesis. (A) In C_3 plants, the rate of photosynthesis reaches a maximum at a lower light intensity than in the case of C_4 plants. (B) In C_3 plants, photosynthetic efficiency decreases as temperature increases, whereas in C_4 plants, efficiency remains constant over a temperature range from 10 °C to 40 °C (after Stern, 2006).

- Plants have evolved to carry out these vital functions in the biosphere with the highest possible efficiency in all the environments on Earth, from the hottest to the coldest to the driest through the evolution of two main, distinct forms of photosynthesis: C_3 and C_4 (Figure 1).

Little wonder, then, that the question of how plants carry out photosynthesis has been one of our major preoccupations for a long time. Its importance to all life ensures that it will continue to engage and fascinate far into the future.

2 Plant respiration: breathing without lungs

INTRODUCTION

An impressive achievement in biology in the twentieth century was gaining a comprehensive understanding of respiration. All living things respire. Still, the knowledge that most people have of what is involved often begins and ends with: "we inhale air rich in oxygen and exhale it enriched with carbon dioxide." But there is much more to it than that.

The foods we eat are slowly burned or "combusted," as Lavoisier described it more than 250 years ago (see Chapter 1). Using the O_2 from the air, we slowly convert carbohydrates, fats, proteins, and other substances, finally, to CO_2 and water. This releases the energy contained in foods, some of it in the form of chemical energy, which is useful to us; the rest is given off to our surroundings as heat. We put the useful energy to work to sustain our life support systems – to drive our muscles and other organs, keep us warm, feed our brains, and build our complex molecules (Figure 2).

PLANTS NEED RESPIRATION AS WELL AS PHOTOSYNTHESIS

We might suppose that because plants have access to an endless supply of energy from the sun they do not need any other source – not so. Not all parts of a plant photosynthesize, only those that are green; however, non-green parts also need energy. In addition, photosynthesis occurs only during the day but plants grow round the clock.

Respiration provides the additional energy needed to build the many types of molecules that plants need for their life processes. Fats, proteins, and nucleic acids are obvious ones, which plants have in common with other organisms. But plants also produce a vast

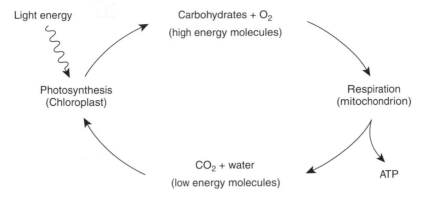

FIGURE 2 The flow of energy in a plant. Plants use the energy of sunlight to manufacture energy-rich carbohydrates from the low energy molecules, CO_2 and water. Mostly in mitochondria of cells, these products of photosynthesis are broken down again to CO_2 and water, releasing energy, some of which is captured as ATP and used to drive the activities of the organism (after Raven *et al.*, 1999).

array of molecules they use for such things as attracting pollinators and for their defense (see Parts III and IV).

PLANT RESPIRATION: A BRIEF HISTORY

Knowledge of respiration in plants had its beginning in the seventeenth century, when it was discovered that seeds must be exposed to air to germinate. But it was not until the work of Lavoisier, Priestley, and others about a century later (see Chapter 1) that the exchange of O_2 and CO_2 between organisms and their surroundings began to be appreciated. Up until that time air was both invisible and mysterious.

A quarter century of progress

Much progress was made during the last quarter of the eighteenth century. For example:

- seeds were shown to take in O_2 and release CO_2 during germination;
- **Ingen-Housz** (see Chapter 1) demonstrated that all living plants give off CO_2 in the dark and that non-green parts of plants do so in the light as well.

What the green parts of plants do in the light was much more difficult to figure out, as we shall see shortly. It was left to de Saussure (see Chapter 1) to undertake the first detailed investigation of plant respiration.

- In the last decade of the eighteenth century he showed that both plants and animals produced and released CO_2 in similar amounts
- He later showed that all parts of a plant exchange O_2 and CO_2 with the surrounding air.

Gas exchange by plants in the light: a recipe for confusion
De Saussure also measured how much O_2 or CO_2 was taken in or given out in the light compared to the dark in green plants. Measuring plant respiration in green tissues in the light has always been a challenge. For example, many plants do not give off CO_2 during the day from these tissues because what they produce in respiration is used at once in photosynthesis. Likewise, some of the O_2 released in photosynthesis is quickly recycled into respiration.

Separating respiration from photosynthesis by measuring exchange of gases by leaves during the day is difficult even with modern instruments – in de Saussure's day the task was impossible.

An enigma: photorespiration
Even when CO_2 *is* given off from green tissues in light, it may originate in **photorespiration**, which has nothing to do with true respiration, but with photosynthesis. Photorespiration provides no energy to the plant but causes some of the CO_2 destined for photosynthesis to be released again before it can be used in forming sugars. In effect, it is an inefficiency in the use of CO_2 in photosynthesis (but see Chapters 12 and 16 for other uses of photorespiration).

Progress stalled
Little progress in knowledge of respiration in plants occurred in the first half of the nineteenth century because the exchange of gases for photosynthesis and respiration continued to be confused with

one another. Not until the latter half of the nineteenth century was a start made on resolving this confusion and true respiration (O_2 in; CO_2 out) separated in plants from photosynthesis (CO_2 in; O_2 out). Knowledge of photorespiration came later still.

RESPIRATION AND ITS COMPLEXITIES

Respiration, like photosynthesis, proved to be far more complex than anyone imagined. There are approximately 50 steps if all aspects of respiration are taken into account. In an attempt to make matters simpler, most explanations of respiration are limited to how glucose is broken down to release energy. The series of steps leading from glucose to the final release of CO_2, water, and energy is much the same in all organisms.

We should remember, however, that in many living things, not only carbohydrates are "burned" in respiration but fats and proteins as well. This is particularly true of those animals where a highly active lifestyle places severe demands on their energy supply systems.

What about plant respiration?

In the vast majority of plants, carbohydrates are by far the main source for respiration. Proteins are not routinely burned for their energy, perhaps because plants produce such large amounts of carbohydrate and have difficulty finding in their surroundings the nitrogen needed for protein formation, which they therefore conserve (see Chapter 5 and Part V).

On the other hand, plant fats are used to generate energy, especially in some seeds (e.g. soybean, canola, mustard) where droplets of oil are stored and are then consumed during germination. On a weight-to-weight basis, fats pack more of an energy punch than do carbohydrates.

Why is respiration so complex?

It is easy to release the huge amount of energy contained in sugars. Throw a pinch of glucose or sucrose onto an open fire, for example, and a hot, bright flame will flare up. But this way of releasing energy as heat, all at one time, is not useful to living things. Occasionally, an organism may direct much of the energy released in respiration to the production of heat (see some examples later in this chapter), but conditions inside cells are a far cry from the uncontrolled conflagrations caused when sugar is thrown into an open flame.

THE RESPIRATORY PATHWAY

The main function of respiration is to trap as much as possible of the energy given off in the "burning" of chemical fuels in a form of greatest value to the organism; in a controlled conflagration. This is what the 50 or so steps in respiration achieve.

- Larger molecules, such as carbohydrates, are broken down to smaller molecules like CO_2 and water
- In the process, some of the energy released, about 40%, is retained in a form that can be used by the organism wherever an input of energy is required
- The rest of the energy released in respiration, some 60%, is given off to the environment as heat in all organisms.

In plants, the greatest demand for energy comes from areas where new growth is occurring, such as at root and shoot tips, and where flowers, fruits, and seeds are being formed.

Glucose and aerobic respiration

Choosing the breakdown of glucose as an example, the main pathway of respiration can be divided into two parts.

First, glucose is *partially* broken down in a series of about nine steps, called **glycolysis**; only a little energy is released and *no oxygen is needed*. Even so, this is an important set of steps which may date back billions of years to a time when there was no O_2 in our atmosphere (see Part V). Today, we see remnants of that more restricted lifestyle in the process of fermentation, more of which in a moment.

In the second part of respiration, the breakdown of glucose is continued all the way to CO_2 and water, a process *requiring* O_2 and releasing a great deal of energy.

Glucose and anaerobic respiration

Some organisms live under conditions where O_2 is either not available or is severely restricted. This can be true of some green plants under conditions such as waterlogged soil or in water bodies where oxygen has been depleted. Some seeds, when they are just beginning to germinate inside thick, hard seed coats, grow under conditions of restricted oxygen until the young seedlings emerge from the seed coats.

Under these relatively *anaerobic* conditions the breakdown of glucose follows the same path as in the first part of the aerobic process where a small amount of useful energy is released. But instead of being converted to CO_2 and water, the partially broken down glucose is converted to ethanol in **fermentation**.

ATP: the universal energy carrier

A single high energy molecule called adenosine triphosphate (ATP for short) has been adopted universally within the living world to deliver the useful energy released during both aerobic and anaerobic respiration to everywhere it is needed (see Box 2).

BOX 2. ADENOSINE TRIPHOSPHATE (ATP): THE UNIVERSAL ENERGY CARRIER

*Early investigations of the role of ATP in living cells were spear-headed by the Nobel laureate, **Otto Fritz Meyerhof**, and his group at the University of Heidelberg. In 1929, **Karl Lohmann** discovered ATP, just one of many phosphate-containing compounds under study by Meyerhof's team. Further research carried out in Heidelberg and elsewhere in the 1930s drew attention to the role of certain types of these molecules in energy release during muscle contraction, including ATP. Using the same calorimetric methods (the calorimeter measures the heat given off when energy is expended), Meyerhof and others then went on to determine, more broadly, that phosphates isolated from living tissues could be divided into two*

BOX 2. (cont.)

groups depending on how much heat was released when they were hydrolyzed (i.e. when a phosphate group was removed by treatment with acid). In most cases, the amount of heat given off was between about -8 and -16 kJ mol^{-1}; others had much more negative values of -30 to -50 kJ mol^{-1}. ATP fell into the latter category with an energy release of about -30 kJ mol^{-1}.

The concept that there were two types of phosphate bond (low and high energy) in phosphate-containing molecules was developed by **Fritz Lipmann** in the early 1940s. In one type, linkage between phosphate and the rest of the molecule was strong, phosphate was liberated only with difficulty, and the amount of energy released was low. Compounds of this sort were called **low energy** phosphates and the bond linking phosphate was a **low energy phosphate bond** (represented as -P). In the other type, the bond linkage was weak, the phosphate more easily released and the energy liberated was high. Phosphate linkages here are **energy-rich bonds** (represented as ~P). By this convention, ATP was written as adenosine-P~P~P to indicate the presence of two energy-rich bonds and one energy-poor bond. On removal of the terminal phosphate group from ATP, adenosine diphosphate (ADP) is formed and much energy is released. Conversely, with an input of energy, an inorganic phosphate group can be made to bind to ADP to form ATP.

ATP acts as the carrier of energy in all organisms. It captures the chemical energy released in combustion of molecules in respiration or, in green organisms, from the absorption of light energy in photosynthesis, and transfers it to reactions that require energy, e.g. the building of cell components, muscle and nerve action, or formation of sugars in leaves. Most of the ATP synthesis is carried out by the enzyme ATP synthase either in the mitochondrion during respiration or the chloroplast during photosythesis.

Surprising quantities of ATP are formed and consumed in organisms. For example, at rest, an adult converts daily a quantity of ATP equivalent to about one-half body weight.

Glucose fermentation to ethanol

The production of ethanol has been used in wine- and beer-making for centuries. In fact, it was German scientists in these industries who first determined the steps in glucose fermentation.

The process requires no oxygen (is anaerobic), releases ethanol and carbon dioxide (the bubbles in wine and beer), gives rise to some energy useful to the organism (ATP), and produces heat. But fermentation releases only about 7% of the total energy stored in glucose; the rest remains locked in the ethanol.

Anaerobic versus aerobic respiration

Although fermentation can produce significant amounts of energy in many microbes, including the yeasts used in wine- and beer-making, it is not nearly efficient enough to sustain the more vigorous lifestyles of green plants, even less so of mobile organisms like animals. **Aerobic respiration**, which takes place in the mitochondria of the cell, is a necessity in these kinds of organisms. Aerobic respiration releases nearly 20 times as much energy as does its anaerobic counterpart.

Thus, most of the green plants that find themselves in conditions where oxygen is restricted (as in waterlogged soil, for instance) cannot survive for very long on anaerobic respiration alone unless they are adapted for living in environments where oxygen is normally in short supply. Not only is there not enough energy being released to maintain most plants under such conditions, but the ethanol accumulated in waterlogged roots is toxic and will kill the root tissues if allowed to build up. Even the most efficient yeasts specifically bred for use in brewing and wine-making cannot survive more than about 23% alcohol in their surroundings. Most green plant tissues, and yeasts for that matter, are killed by much lower concentrations.

CYANIDE-RESISTANT RESPIRATION

In many organisms, including some plants, aerobic respiration is quickly poisoned by cyanide. In many plants, however, respiration continues for some time in the presence of cyanide. This so-called **cyanide-resistant** respiration is found also in some fungi, algae, and a few animals.

Cyanide stops respiration by blocking steps in which oxygen is used and which lead to energy release. In many living things that depend on aerobic respiration this quickly leads to death.

However, many plants have an alternative route for using oxygen and releasing energy in respiration if cyanide is present; most of the energy in these cases is released as heat instead of to the energy carrier ATP.

Practical uses of cyanide-resistant respiration
In thermogenesis

In arum lilies, like skunk cabbage, heat is used in their pollination strategy. When the arum lily flower is ready for pollination, its temperature can rise as high as 30 °C, which causes certain chemicals in the flower to evaporate. The scent released into the air attracts potential pollinators (see Chapter 11).

Certain Arctic plants create cosy spots for insects by stepping up flower temperature by as much as 8 °C. In this way, the Mountain Avens and Arctic Poppy, for example, make themselves attractively warm for pollinators.

Many plants, such as crocuses, which bloom very early in the spring in the northern hemisphere, produce enough extra warmth to help melt snow in their immediate surroundings by raising their temperature by a few degrees.

The idea of the use of the alternative respiratory pathway to release heat is known as the *thermogenesis hypothesis*.

In energy overflow

Although this alternative pathway of respiration has its practical uses, cyanide will kill plants eventually. Plants, like all other organisms, cannot live by heat energy alone.

More controversial is what role the alternative pathway has under normal conditions. Many more plants have the pathway than use it for tasks such as raising their temperature to aid in pollination or to melt snow. For what do these plants use it?

The alternative pathway has its highest activity in plants rich in carbohydrates, for example after rapid photosynthesis.

Some have suggested that the speed of their breakdown when carbohydrates are in abundant supply might overwhelm the normal route of energy release in respiration and that the alternative pathway acts as an overflow breakdown pathway to relieve pressure on normal processes. Perhaps plants are not able to control their release of energy in respiration in any more efficient way than this, a notion which has become known as the *energy overflow hypothesis*.

OTHER USES OF THE RESPIRATORY CHAIN

It would be wrong to leave the impression that respiration begins and ends with the release of energy whether that be in a form that the plant can use to help drive its life functions or as heat. At each of the 50 steps in the chain, the metabolites formed can do one of two things.

Each step in the chain can simply lead to the next, moving inexorably towards the consumption of oxygen and the release of CO_2, water, and energy.

Alternatively, the plant can siphon off a percentage of each chemical formed for other uses, returning it to the chain later. Products from the respiratory pathways are used in the manufacture of thousands of new compounds in plants. For example, waxes on leaves, colored pigments like chlorophylls and carotenoids, terpenes such as rubber, and hormones like the gibberellins, are a single, huge grouping of compounds all formed from the same source in respiration. Fats and oils also arise from this same source and can be converted back into carbohydrates.

Chemicals produced at other points in the pathway can be redirected to form amino acids needed for protein formation, the building blocks of the genetic material deoxyribonucleic acid (DNA) and ribonucleic acid (RNA), and other important molecules such as hormones like auxins, gibberellins, cytokinins, ethylene, and abscisic acid (ABA), as well as the alkaloids, such as caffeine and nicotine, and many more.

RATE OF RESPIRATION

It would also be wrong to leave the impression that respiration proceeds in all plants and in all parts of a plant inexorably, at the same speed, day and night. Many factors influence the rate of plant respiration, which can vary just as it does in animals like ourselves. The more vigorously we exercise, for example, the higher our rate of respiration. During sleep our respiration slows down.

Day versus night

Starved plants low in starch or stored sugars respire more slowly but speed up again when supplied with carbohydrate. Respiration in leaves is often fastest just after sundown when carbohydrate levels are high after the day's photosynthesis; it is lowest just before sunrise.

Shade versus sun

Shaded leaves low on a plant often have a slower rate of respiration than leaves further up the stem where they are exposed to higher light levels and where, therefore, more carbohydrate is formed.

Protein respiration

In extreme starvation conditions even proteins can be respired, but this is a last resort in plants as it is in animals.

People on hunger strikes use up the carbohydrate and fat reserves in their bodies before there is significant breakdown of proteins. After all, the majority of proteins in the animal body are in the form of muscle tissue and the thousands of enzymes needed to keep the machinery of the body working.

In plants, maintaining muscle tissue is not an issue but enzymes are. Breaking down enzymes to provide energy makes little sense, however, since without these catalysts energy is useless. As a last resort, some less essential proteins may be broken down to provide energy for crucial core life processes.

Fruit ripening, the climacteric

Variations in rate of respiration also occur during development of ripening fruits.

In all fruits, the respiratory rate is high when they are young and growing rapidly, declining as the fruit matures.

In many species this gradual decline is reversed by a sharp increase at the time of approaching full ripeness and flavor of the fruit (see Chapter 14, Figure 14).

Common fruits showing this spike in respiration, called the **climacteric rise**, include apples, bananas, pears, peaches, nectarines, and tomatoes; citrus fruits, like oranges, lemons, and grapefruits, as well as cherries, grapes, pineapples, and strawberries, do not.

Why some fruits have the climacteric and others don't is unclear. What is clear is that the sudden rise in respiration in certain fruits is preceded and triggered by a sharp increase in production of the hormone, ethylene.

SENSITIVITY OF PLANTS TO OXYGEN SUPPLY

As already discussed above, fermentation occurs in conditions where oxygen is either not available or is in short supply. This raises the question as to how sensitive plants are to the presence or absence of oxygen in their surroundings and how they ensure delivery of an adequate supply of oxygen for respiration to all tissues.

Slight variations in oxygen concentration in normal air probably have no measurable effect on plants. There is also generally no problem in maintaining a constant, steady, high level of oxygen in most leaves, stems, and roots.

In bulkier tissues

Bulkier tissues, such as carrots, potato tubers, and other storage organs, have lower rates of respiration at their centers than closer to their surfaces. However, there is still enough oxygen in the air deep in these dense organs for respiration to be aerobic, not anaerobic.

Air spaces help

Air spaces are important in storage tissues as well as in organs that normally grow in conditions where oxygen is less freely available. In potato tubers, for example, about 1% of the tissue volume is occupied by air spaces. In roots from a wide range of species, air spaces can be seen to occupy between 2 and 45% of the root volume. The higher values are common among plants growing in wetlands where wide air tubes or spaces may extend all the way from the leaves into the roots. Air taken in through stomata in leaves is moved efficiently down to roots which may be immersed in water where oxygen is depleted.

ADAPTATION TO OXYGEN DEPLETION

In grasses and sedges

Some plants are better adapted than others to withstand oxygen depletion.

Grasses and sedges have hollow stems through which air can be supplied to roots even under flood conditions. Among crop plants, only rice is known to tolerate oxygen depletion for long, partly because of the fact that rice seeds can germinate in water, where the oxygen level is low, by relying on efficient fermentation to supply energy. Young rice seedlings take this energy and channel it into rapid growth that thrusts the seedling swiftly above the water surface. Once under aerobic conditions, air can be funneled directly to the roots still submerged in water through wide air tubes extending the length of the plant.

In bulkier plants

Even large plants, like shrubs and trees, have different tolerances to oxygen availability.

In tropical mangroves some roots grow vertically out of the water in which the trees live more or less continuously. Air is transferred down through these specialized organs to other submerged root tissues.

Among conifers, lodgepole pine is more tolerant to flooding than is the Sitka spruce. The main difference, again, is the greater ability of the pine to transport oxygen down to its roots.

SUMMARY

There is much more to respiration than just simple exchange of oxygen and carbon dioxide.

- Larger molecules, such as the carbohydrates, fats and, sometimes, proteins, are oxidized to carbon dioxide and water in aerobic respiration
- When oxygen is unavailable or is restricted in supply, other endpoints than the formation of carbon dioxide and water occur in anaerobic respiration
- Fermentation, the formation of ethanol, is one anaerobic pathway many organisms, plants included, use to generate the energy they need. Fermentation, however, is a much less efficient route of energy release than its aerobic counterpart and cannot sustain organisms as large and vigorous as most green plants
- As an alternative, some plants have ways of moving air to areas where oxygen is restricted, thus favoring aerobic respiration. Air tubes or spaces within tissues allow air to be transported from the aerial parts of a plant to roots where there is less oxygen available
- In both aerobic and anaerobic respiration, the molecule ATP is the preferred carrier of the fraction of energy which can be used by the plant for life support; the rest is released as heat
- Cyanide-resistant respiration is another route of energy release common in the plant world although not unique to it. Here, energy is released mainly in the form of heat which has limited value to a plant. Some plants use the warmth produced in this type of respiration for particular tasks but, in most cases, it may be an overflow mechanism linked to regular respiration

- Respiration is not simply a way to generate energy. Along the pathways of respiration, substances are formed which are the foundation of branch routes leading to thousands of other organic compounds
- In rapidly growing numbers of cases these substances are known to play crucial roles in the life of the plant
- In thousands of other cases, roles are only now beginning to be understood through the many advances being made in the techniques and technologies of molecular biology
- In yet other instances, the reasons for their formation are still mysterious.

GENERAL REFERENCES: PARTS I–IV

Hopkins, W. G. and Hüner, N. P. A. (2009). *Introduction to Plant Physiology*, 4th edn. Hoboken, NJ: Wiley.

Huxley, A. (1978). *Plant and Planet*. Harmondsworth, Middlesex: Penguin.

Raven, P. H., Evert, R. E. and Eichhorn, S. E. (1999). *Biology of Plants*, 6th edn. New York: Freeman.

Ridge, I. (ed.) (2002). *Plants*. Oxford: Oxford University Press.

Salisbury, F. B. and Ross, C. W. (1992). *Plant Physiology*, 4th edn. Belmont, CA: Wadsworth.

Stern, K. R. (2006). *Introductory Plant Biology*, 10th edn. New York: McGraw-Hill.

Taiz, L. and Zeiger, E. (2006). *Plant Physiology*, 4th edn. Sunderland, MA: Sinauer.

BIBLIOGRAPHY: PART I

Bazzaz, F. A. and Fajer, E. D. (1992). Plant life in a CO_2-rich world. *Scientific American*, **266**, 68–74. *That possible benefits of increasing CO_2 in the atmosphere are rooted in the details of the pathway of photosynthesis, an issue also discussed at length in Part V.*

Bjorkman, O. and Berry, J. (1973). High-efficiency photosynthesis. *Scientific American*, **299**, 80–93. *Authors consider the range of environments in which plants grow and the adaptations in photosynthesis that have evolved in these environments.*

Boyer, P. D., Walker, J. E. and Skou, J. C. (1997). ATP – the universal energy carrier in the living cell. *Press release from the Royal Swedish Academy of Sciences, 15 October 1997, announcing the Nobel Prize in Chemistry.* http://nobelprize.org (accessed July 10, 2009).

Castelvecchi, D. (2009). Photosynthesis: the reaction that makes the world green is just one of many variants. *Scientific American*, **301**, 89. *A note regarding the type of photosynthesis in which water splitting is used to generate a source of chemical energy.*

Dostrovsky, I. (1991). Chemical fuels from the sun. *Scientific American*, **265**, 102–7. *Shows how much energy the sun delivers to our planet and how this source might be harnessed by humans.*

Engelmann, Th. W. (1882). On the production of oxygen by plants in a microspectrum. *Botanische Zeitung*, **40**, 419–26. In *Great Experiments in Biology*, M. L. Gabriel and S. Fogel (eds). Englewood Cliffs, NJ: Prentice-Hall, 1955. *Here, Engelmann used the alga* Cladophora *and bacteria to answer the question of which wavelengths of light were most effective in photosynthesis.*

Hill, R. (1939). Oxygen produced by isolated chloroplasts. *Proc Roy Soc Lond, Ser B*, **127**, 192–210. In *Great Experiments in Biology*, M. L. Gabriel and S. Fogel (eds). Englewood Cliffs, NJ: Prentice-Hall, 1955. *In which Hill describes how oxygen release in photosynthesis is linked to the light and can occur even when chloroplasts are isolated from cells.*

Ingen-Housz, J. (1779). *Experiments upon Vegetables, Discovering that their Great Powers of Purifying the Common Air in the Sunshine, and of Injuring it in the Shade and at Night.* London: P. Elmsly and H. Payne. In *Great Experiments in Biology*, M. L. Gabriel and S. Fogel (eds). Englewood Cliffs, NJ: Prentice-Hall, 1955. *Ingen-Housz showed that oxygen was produced only in the light in photosynthesis and only in the green parts of plants.*

Khakh, B. S. and Burnstock, G. (2009). The double life of ATP. *Scientific American*, **301**, 84–92. *ATP is being recognized now as not just an essential source of energy in cells but also as a carrier of critical messages between cells.*

Kiang, N. Y. (2008). The color of plants on other worlds. *Scientific American*, **298**, 48–55. *Do plants need to be green? The author reasons that plants could be red, blue, or even black.*

Lambers, H. and Ribas-Carbo, M. (eds) (2005). Plant respiration: from cell to ecosystem. *Advances in Photosynthesis and Respiration*, vol. 18. Dordrecht: Springer. *Comprehensive coverage of most aspects of knowledge about respiration in plants, including photorespiration.*

Maruyama, K. (1991). The discovery of adenosine triphosphate and the establishment of its structure. *J Hist Biol*, **24**, 145–54. *Concerning the initial discovery and establishment of the structure of ATP.*

Priestley, J. (1772). Observations on different kinds of air. *Phil Trans Roy Soc Lond*, **62**, 166–70. In *Great Experiments in Biology*, M. L. Gabriel and S. Fogel (eds). Englewood Cliffs: NJ: Prentice-Hall, 1955. *Shows that green plants "restore" air.*

Ruben, S., Kamen, M. and Hyde, J. L. (1941). Heavy oxygen (^{18}O) as a tracer in the study of photosynthesis. *J Am Chem Soc*, **63**, 877–9. In *Great Experiments in Biology*, M. L. Gabriel and S. Fogel (eds). Englewood Cliffs, NJ: Prentice-Hall, 1955. *Shows that oxygen released during photosynthesis comes from water, not carbon dioxide.*

Saussure, N.Th. de. (1804). On the influence of carbonic acid gas on mature plants. *Recherches Chimique sur la Vegetation, Paris.* In *Great Experiments in Biology*, M. L. Gabriel and S. Fogel (eds). Englewood Cliffs, NJ: Prentice-Hall, 1955. *Showing that CO_2 from the air is required for photosynthesis.*

Staes, D. M. Otto Meyerhof and the Physiology Institute. (2001). The birth of modern bio-chemistry. In *A History of the Kaiser Wilhelm Institute for Medical Research* 1929–1939, Nobelprize.org, 2001. *A report from the Nobel Prize Organization on the career of Otto Meyerhof who, along with A. V. Hill, was awarded the Nobel Prize for Physiology and Medicine in 1922.* http://nobelprize.org

Stiles, W. and Leach, W. (1952). *Respiration in Plants*, 3rd edn. London: Methuen and Co Ltd. *Detailed account of early history of respiration from de Saussure onwards.*

Van Helmont, J-B. (1748). By experiment, that all vegetable matter is totally and materially of water alone. *Ortis Medicinae*, 108–9. In *Great Experiments in Biology*, M. L. Gabriel and S. Fogel (eds). Englewood Cliffs, NJ: Prentice-Hall, 1955. *Translation of Van Helmont's great experiment with willow cuttings by N. Lewis.*

Van Niel, C. B. (1929). *Contributions to Marine Biology*, Stanford: Stanford University Press, 161–9. In *Great Experiments in Biology*, M. L. Gabriel and S. Fogel (eds). Englewood Cliffs, NJ: Prentice-Hall, 1955. *Van Niel clarifies the nature of variants of photosynthesis among which is the notion that oxygen comes from the decomposition of water.*

Part II Plant nutrition

Plants must gain from the environment the raw materials they need to sustain their structure, growth, and development. In Part I, we learned how they use light energy in photosynthesis and chemical energy generated through respiration to manufacture a wide array of organic chemical compounds. The focus in Part I was very much on the role of carbon from carbon dioxide in this endeavor. Part II is devoted to the question of what else plants need to sustain their growth and development.

In contrast to animals, the nutritional requirements of plants are simple. In addition to carbon dioxide, they need only water and certain chemical elements, all of which they most often take up from the soil. In the first two chapters in Part II, we examine how plants take in and distribute the **water** and **mineral elements** they require. The third chapter deals separately with the special case of how plants acquire and use **nitrogen**. A final chapter describes how plants **transport** substances of all kinds within and between their various organs.

These physiological processes are often grouped under the collective heading of **plant nutrition**.

3   Plants are cool, but why?

INTRODUCTION

Plants are able to shed surplus heatloads in a number of ways, two
of which are especially important but only one of which leads to the
coolness we associate with places where plants are abundant, such
as forests and meadows.

CONVECTIVE VERSUS EVAPORATIVE LOSS OF HEAT

If the temperature of a leaf is higher than its surroundings, air circulation will remove heat from its surface mainly by **convection**. As
this warm air rises, it cools, becomes more dense, and sinks, creating a convection current which removes heat from plant surfaces.

Evaporation of water from leaf surfaces withdraws heat from
a plant because energy is absorbed by water as it changes from liquid to vapor (the latent heat of vaporization of water is 44 kJ mol^{-1}).
Evaporative cooling can occur even if the temperature of the leaf is
below that of the surrounding air.

Which of these two ways of heat loss is the most important to
a plant depends on its environment. If there is an ample supply of
water then loss of it from leaves can be high without causing damage to the plant; evaporation can be a major means of cooling. Plants
adapted to growing in hot, dry conditions, on the other hand, have
evolved ways to conserve rather than shed water; convective air currents become the main route for shedding heat.

A typical leaf at moderate temperature dissipates about half its
heatload by evaporation of water, half by convection. Plants with a
bias to convective heat loss often have thick, fleshy tissues in which
water is stored. They can often withstand temperatures above 50°C
without suffering damage; cacti are a good example. Those with a

bias towards an evaporative water strategy for shedding heat have evolved to take up water from the soil rapidly, transport it swiftly upwards to leaves, and evaporate it into the air, efficiently.

PLANTS ALSO NEED WATER

Although plants may evaporate away as much as 98% of all the water passing through them in a lifetime, we must remember that they also need it in photosynthesis, as a medium in which to carry out chemical reactions in their cells, to keep living cells firm (i.e. maintain their turgor) for functioning, and to prevent wilting of tissues and organs. In fact, water comprises 70–95% of a plant's weight.

WATER UPTAKE AND DISTRIBUTION BY PLANTS

During its lifetime, a plant growing in temperate regions of the world may lose the equivalent of 100 times or more its own weight of water by evaporation. In some tropical regions, losses may be higher yet. In hot or cold deserts, where plants conserve water, losses are far lower.

Most plants, regardless of their environments, have efficient ways to bring in water from their surroundings. The main way is usually the soil via the roots but some plants can take water in directly from the air. In tropical rain forests, for example, where the humidity is high, aerial roots growing out from branches absorb water directly from the air.

THE ROLE OF ROOTS

The size and reach of root systems

Root systems can be enormous compared to plant growth above ground. In one case, a mature rye plant growing under ideal conditions was found to have a total root length of 622 km with an average growth rate of 5 km a day, and a total root hair length of 10 620 km with an average growth rate of 90 km a day.

The staggering quantity of root hairs on a single plant hints at why, when plants are first transplanted, they grow less well until better established. When a plant is disturbed, many of its delicate root hairs are stripped away and its roots function poorly until new hairs are formed. Commonly, a root tip may have around it some 2500 hairs cm^{-2}, which increases the absorbing surface of the root by up to 20-fold.

Roots extend continuously during a growing season into new areas of the soil, searching out water and, as we shall see in the next chapter, also the minerals dissolved in it. This constant quest for water and nutrients takes the root system often far down into the soil. Roots at depths of 1–2 metres are common but can be found at more than 50 metres in some desert species. They can spread sideways as well as downwards and become very closely associated with just about every particle of a large volume of soil within what is often called the **rhizosphere**.

Root systems enhanced by soil fungi
Most, if not all, roots have associated with them a wide array of microbes which enhance their capacity to absorb water and nutrients from the soil. The most widespread and, from a nutritional perspective, the more significant associations are those between roots and soil fungi.

The formation of a **mycorrhiza** (fungus root) is typical of more than 80% of roots, a reflection of the mutual importance these associations have for both the plant and the fungus. In some cases, the plant root and surrounding fungi are only loosely linked to one another; in others, the fungi penetrate into the root itself, as well as extending out into the soil.

MOVEMENT OF WATER IN THE PLANT
Once inside the root, water must be moved to all parts of the plant body, especially to the leaves where it exits to the surrounding air. Some plants grow to impressive heights. For example, the giant redwood trees of California are impressively tall but some Douglas

firs of the Pacific Northwest of North America and eucalypts of Australia are among the tallest plants in the world, at 120–150 m in height. To reach the highest leaves of such behemoths, water must rise from roots below ground level a vertical distance greater than this. Thus, any believable explanation of how water moves up a plant must account for several important points.

Pressure problems

Just to raise water a distance of, say, 130 m above the ground requires a push from below, or a pull from above, of 1.3 MPa. To move water upwards this distance against the resistance of the dense tissues of a plant may require a force of 3.0 MPa. By way of comparison, a car tyre is inflated to a pressure of only about 0.2 MPa (two atmospheres or 30 lbs per square inch). A pressure of 0.3 MPa is about the point at which a diver begins to suffer the first signs of respiratory discomfort, a depth of around 30 m in water. To experience 3.0 MPa, a diver would have to swim to a depth of 300 m, a feat which would cause crushing problems within the body.

Problems of speed and volume

Any explanation of the rise of water up a plant must also account for the speed and volume of water moved. In some hardwood trees, for example, water may rise at the rate of almost 50 m an hour. A full grown maple tree in open country may lose more than 200 L of water in an hour on a warm sunny day. Measurements made under the hot conditions found in the mid-west of the USA showed that a maize plant may evaporate away as much as 200 L of water in a lifetime, approximately 100 times its own mass.

Plant structure and water movement

Finally, an explanation of plant water movement must be consistent with the structure of plants. The upward flow of water is known to take place in the woody tissue. Dead cells in the wood (the **xylem**) act as microscopic tubes through which water, and anything dissolved in it, moves from the roots to the furthest branches and leaves. These xylem tubes (**tracheids** and **vessels**) form a continuous system of very narrow but open pipes running the entire length of the plant as well as into all tissues along the way. The plant must have a mechanism to move sap through these open pipes at a sufficiently high rate to satisfy all a plant's needs.

RISE OF WATER EXPLAINED
The first coherent explanation of how water rises in plants was put forward nearly 300 years ago.

Stephen Hales and staticks
In 1727, **Stephen Hales**, a versatile English clergyman and scientist, published a highly influential book entitled *Vegetable Staticks*. For this and other foundational studies, Hales is rightly regarded as the "father of plant physiology." A good portion of his book is devoted to the rise of water in plants, which Hales investigated by weighing (the science of staticks) plants before, during, and after various treatments.

Of the two ways in which plants do move water over long distances, Hales did considerable work on one and hinted at the other.

Root pressure
The idea of **root pressure** is an old one investigated in great detail by Hales. He found that plant roots sometimes develop a pressure build-up owing to their ability to absorb water rapidly from the soil, and suggested that this pressure accounted for the rise of sap in a stem.

The ability of plants to take in water rapidly from the soil is based on their capacity for **osmosis**, a process of great importance to all living organisms.

Osmosis and cell membranes
Osmosis depends on the fact that each living cell of an organism is surrounded by a delicate membrane so fine that it cannot be seen

with the naked eye. Despite its apparent flimsiness, this barrier is of enormous importance. It not only separates what is inside each living cell from what is outside but also controls exchange of the substances that are constantly moving between the interior of a cell and its environment. Salts and other water-soluble compounds contained within the living cell are held in by the membrane and allowed to move across it only slowly. This strong intracellular solution attracts water from the surroundings into the cell.

Aquaporins

Thus osmosis is the attraction of water across a controlling barrier, such as a living cell membrane, in the direction of a stronger solution of dissolved substances. Water can move across a cell membrane more freely than the solutes dissolved in it because of the presence of pores dedicated to the rapid transit of water. These **aquaporins** are protein-controlled pores which can be opened or closed depending on a cell's need for water (Figure 3).

All membranes, plant or otherwise, have **porins**, a family of proteins which control the movement of many substances into and out of cells or from one cell compartment to another (e.g. mitochondria, chloroplasts, cell vacuoles), including water.

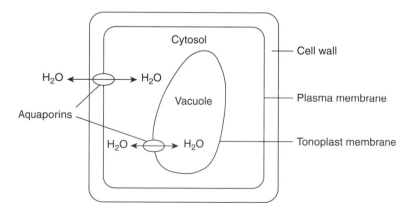

FIGURE 3 Aquaporin channels are proteins found in plasma and tonoplast membranes of living cells. They can be opened and closed to regulate water flow (after Hopkins and Hüner, 2009).

Turgor pressure and the cell wall

The net result of the accumulation of water is the build-up of pressure within a cell, referred to as **turgor pressure**. The effect is similar to what happens when a balloon is filled with air. The pressure inside the balloon increases until no more air can be pumped in without the balloon bursting. In the case of the living cells of a plant root or root hair, the build-up of turgor pressure caused by the extraction of water from the soil does not lead to the bursting of cells because they are surrounded not just by a filmy membrane but also by a tough, cellulose wall. Instead, the pressure build-up can be used to force water from the root into the stem and leaves.

Back to root pressure

Root pressure is well known and significant in some plants: a potent liquor called pulque is made by native Central Americans from the sap of the century plant. By cutting off the single flower bud as soon as it appears, sap can be collected as it drips from the cut, then fermented. Over 4 or 5 months the amount of sap seeping out under pressure from the roots of a single plant can approach 50 kg. It is possible to see droplets of water forced out under pressure from the tips or edges of grass leaves, a phenomenon called **guttation**.

However, root pressure is not measurable in many plants, occurs only at certain times of year in others and, even in the best examples, is not strong enough to account for the pumping of sap over the great vertical distances found in trees. Root pressures in most species do not exceed 0.1 MPa and are never greater than 0.3 MPa, far below the force needed to pump water to the top of a 150-m tall tree.

Eduard Strasburger's massive experiment

Even Hales recognized that root pressure could not be the full explanation for water movement within a plant. Others more recently have agreed with him.

In one famous and grand experiment performed in the late nineteenth century, the German botanist, **Eduard Strasburger**, cut off a 22-m tall oak tree close to the ground and immersed the cut-end of the trunk in a bath of picric acid, which kills living cells on contact. The picric acid spread throughout the tree, killing living cells as it progressed. After a few days, Strasburger replaced the picric acid with a bath of water stained with eosin, a red dye, and followed its movement all the way to the leaves even though the picric acid had killed all cells along the transport route.

Thus did Strasburger show that a living pump mechanism was not needed to raise water to the leaves of a tall plant. He demonstrated that even dead tissue could act as a channel through which water could be moved to great heights.

TRANSPIRATIONAL PULL

A hint about what eventually became the main explanation for water transport through a plant body was provided, again, by Hales, who wrote in *Vegetable Staticks*:

> although the [woody tissues of the plant] imbibe moisture plentifully; yet they have little power to protrude it farther, without the assistance of the perspiring leaves, which do greatly promote its progress.

Thus Hales hinted that he understood the role played by evaporation from leaves in the transport of water through the living plant, a process he called "**perspiration**" and we call "**transpiration**."

REQUIREMENTS FOR LONG-DISTANCE TRANSPORT OF WATER

Cohesion and the tensile strength of water
The explanation which seems to satisfy all requirements for rapid, long-distance movement of water through plants is transpiration, the evaporation of water from leaves. The explanation depends entirely on what are known as the **cohesive properties** of water.

Cohesion is a measure of the strength of the chemical bonds holding molecules together; a certain force is needed to tear the bonds apart. The force needed to break the bonds in water, as in any other molecule, is a measure of the **tensile strength** of the liquid.

Among all known liquids, the tensile strength of water is by far the greatest. A thin column of pure water, enclosed in an air-tight tube, free of dissolved gas, can withstand a pulling force of –25 to –30 MPa before the molecules are torn apart, about 10% the tensile strength of copper wire.

The tensile strength of sap

The sap being transported around plants from the roots to the stems and leaves does not need to have tensile strength as great as pure water but has been measured certainly at 20 MPa.

Even a thin column of plant sap enclosed in a column of xylem tracheids or vessels, then, has the cohesion, or tensile strength, to withstand the kinds of forces needed to suck the sap up to a height of nearly 2000 m above the ground without snapping. This degree of cohesive strength is much more than is needed to account for the rise of sap to the furthermost tips of the tallest trees. Tensions in xylem tracheids and vessels vary widely among species and environments, but are typically in the range of –0.5 to –2.5 MPa in temperate deciduous trees.

What creates the sucking force?

The force comes from the creation of a negative pressure similar to how we suck liquid up a straw. First, we suck the air out of the straw, creating a low air pressure inside it. Liquid then moves up the straw to replace the withdrawn air. In a similar way, water lost from leaves by evaporation creates a low pressure within the plant. Replacement water is sucked into leaves along the xylem tubes from the stem and roots and, ultimately, from the soil.

The tensions generated by this sucking action are large. A pull of –2.0 to –3.0 MPa is not unusual in the very tallest plants growing in dry conditions, still well within the known tensile strength of plant sap but well above any recorded force produced by root pressure.

Bubbles are a problem

The large negative pressures found in the xylem tubes of plants when high volumes of water are moved rapidly up to leaves do create problems, one of the most serious of which is that water under tension is unstable (i.e. **metastable**).

As the rate of water movement increases in response to the needs of the plant during daylight hours (mainly for cooling), tension in the water columns in the wood also increases, which causes air dissolved in the water to escape and form gas bubbles in the xylem tubes. Formation of bubbles breaks the continuity of the water columns, halting water transport. Too many breaks of this kind in the tube system, if not repaired, would be disastrous to the plant, causing dehydration and death. Fortunately, at night, when the evaporation of water from leaves is much lower, tensions in the tubes decrease to the point where gas bubbles re-dissolve, restoring the continuity of the columns of sap.

We know how important gas bubbles in plant stems can be. Before placing cuttings (cut flowers, for example) in water we routinely remove the last centimeter or two of the stem, under water. We do this to eliminate any air bubbles that might have found their way into the cut end of the stem and that might impede further uptake of water.

THE DILEMMA OF GASEOUS EXCHANGE

It would be wrong to end here and leave the impression that the transport of water through a plant and out through leaves is an isolated process. Hales, in his *Vegetable Staticks*, made reference to the other major role that leaves play in the life of the plant: photosynthesis and respiration.

> leaves seem also designed for many noble and important services … plants probably drawing thro' their leaves some part of their nourishment from the air.

Most land plants are faced with a major dilemma, in being surrounded by a dry atmosphere into which they transpire water vapor. The air is so far from being saturated with water in most regions of the world that plants would be in danger of lethal dehydration if they were not able to control loss of water.

At the same time they must take in from the air the gases needed for respiration (oxygen) and photosynthesis (carbon dioxide). As these gases enter the leaf, water vapor is lost.

The dilemma facing land plants is how to stop, or at least slow, loss of water without harming themselves by cutting off the flow of O_2 and CO_2 into the leaf.

THE CUTICLE

Most of the above-ground surface of the vast majority of land plants is covered by a thin layer of wax through which gases and water have difficulty passing. We can see this waxy barrier (the **cuticle**) in leaves, which appear shiny and are slippery to the touch. The cuticle serves as an effective barrier to water loss (only 5–10% of water lost by a plant exits via the cuticle), protecting the plant from desiccation as well as from attack by insects and other predators.

However, if this barrier were complete, it would also block intake of O_2 and CO_2. Plants resolve their dilemma in much the same way as do insects, which also have an outer cuticle that is largely impermeable. They, too, are faced with the problem of exchange of air for respiration, a problem they overcome by having breathing pores in the cuticle through which gases are exchanged with the atmosphere.

PLANT BREATHING PORES: STOMATA

Most abundantly on the undersides of their leaves, plants also have breathing pores, called **stomata**. Each **stoma** (the Greek word for "mouth") has two cells on either side of its opening, known as **guard cells**, shaped rather like the lips of a mouth. These two cells are designed so that when they are filled to capacity with water they separate from one another. As they separate, a gap appears between them through which gases can be exchanged between the interior of the leaf and the surrounding air, just as happens when our lips open. At other times, the guard cells lose water, come together, and the stomata close, cutting off gas exchange with the atmosphere.

Control of the movement of water into and out of the guard cells is by osmosis, a process described earlier in this chapter in relation to root pressure.

Guard cell control

The number of stomata in a leaf is normally so great that if they were always open the leaf would evaporate most of its water. The ability to open and shut, however, allows control of water loss.

Pairs of guard cells function as valves with multiple controls to open and close them. Changes in such things as brightness of light, carbon dioxide concentrations inside the leaf, temperature, relative humidity of the air, and water availability from the soil, can all be sensed by guard cells. The cells react by changing their shape, hence altering the size of the opening between them (see Box 3).

At night, when there is no photosynthesis, CO_2 builds up inside the leaf, signaling to the plant that stomatal openings can be kept small, preventing unnecessary loss of water since the heatload on the plant is also small at night; evaporation is not necessary for cooling.

BOX 3. THE STOMA: A MIRACLE OF EVOLUTION

Stomata (pores) in the epidermis of leaves allow exchange of gases between a plant and its environment. Over 90% of CO_2 uptake for photosynthesis and water vapor loss in transpiration occurs through these openings, as does entry of gaseous pollutants such as ozone and sulfur dioxide (see Part V).

The frequency of their occurrence varies widely in a general range from 20 to 400 mm^{-2} of, most often, lower leaf surface. When fully open, a typical stoma measures 5–15 μm in width and 20 μm in length. Combined, stomatal pores comprise only about 1% of

BOX 3. (cont.)

the total leaf surface area. Thus, it might be expected that movement of gases across a leaf epidermis would be severely restricted
since it occurs by diffusion alone, governed by **Fick's Law**:

Rate of diffusion = $D.A(dc/dx)$

where D = the coefficient of diffusion, A = the area over
which diffusion occurs, and dc/dx = the concentration gradient
over which diffusion occurs.

Diffusion is directly proportional to area, A.

However, in the early 1900s, **Horace T. Brown** and
Fergusson Escombe showed that movement of gases through
small pores was proportional to their diameter, not their area;
i.e. it is related to circumference $(2\pi r)$ not area (πr^2). There is a
significant "spill over" effect at the perimeter of small pores,
proportional to the amount of edge. Brown and Escombe used
round pores in their experiments; the elliptical shape of a stoma
exaggerates even further the amount of edge in proportion to
area. As a result, the rate of CO_2 uptake by an actively photosynthesizing leaf can be as high as 70% of the rate expected across
an area equal to that of the entire leaf, many-fold higher than
total stomatal area suggests is possible.

Thus, stomata are exquisitely **shaped** for their role.

Because of the critical importance of water status to a
plant, it is not surprising that stomata are particularly responsive to water availability to the plant in their function in transpiration. There is a highly sensitive "feedback" based on a
continuous evaluation of water status to regulate the degree to
which a stoma is open; fully, partially, or not at all.

Stomata are located at the boundary between the inside
of a leaf, where relative humidity is 100%, and the outside air,
which is usually far less saturated with water vapor. Thus, the
stoma acts as a variable resistance at the exact point where the

BOX 3. (cont.)

concentration gradient of water vapor is at its greatest. Even a slight change in the size of a stomatal aperture increases or decreases the rate of water vapor loss to the outside air.

*Stomata are exquisitely **controlled** to fulfil their critical role in water conservation by the plant (see also Box 12).*

The role of cryptochrome

Even before sunrise, stomata are induced to begin opening through the action of the pigment, **cryptochrome** (see Box 9, Chapter 9). This pigment, found in guard cells, senses the presence of blue light, which is especially abundant in the dawn's early light, causing the stomata to begin opening. As daylight strengthens, the demand inside the leaf for CO_2 for photosynthesis grows, its concentration inside the leaf falls, and stomatal pores continue to open, allowing free exchange of gases.

The role of abscisic acid in plant water conservation

As long as water is plentiful it is to the advantage of the plant to trade water for CO_2 to manufacture food. But when water is not so abundant in the soil, the roots send a signal to the leaves via the xylem water stream. Roots under water stress increase their level of the stress hormone, **abscisic acid** (ABA; see Chapter 7 for discussion of plant growth substances), some of which finds its way into the xylem and, eventually, to the leaves. The arrival of this signal stimulates stomata to remain partially, or even completely, closed.

Of course, restricting stomatal pore size during the day slows photosynthesis but a plant can survive for much longer with a restricted food source than with a shortage of water, just as we can. When was the last time you heard of someone going on a water strike? Hunger strikes are a common means of protest but not water strikes. We can live for weeks without food but not many days without water. Plants are much the same as animals and will conserve water at the cost of making food.

SUMMARY

- We began this chapter by exploring the notion that many plants keep cool by evaporating water from their leaves
- In the course of learning how this happens we also found out that much more is involved in how water is used by plants
- Water is essential to all life. Plants use it in many ways other than in keeping cool and have evolved ways to take it in and move it, often over great distances, against gravity, efficiently, through tubes that extend all the way from the roots to the furthest tips of the highest branches
- Transpiration of water from leaves is the main driving force used to move water from the soil to all parts of the plant
- The entire plant is, thus, supplied, not only with water, but also, incidentally, with minerals dissolved in it (see Chapter 4), all of which are essential to the continued wellbeing of the growing plant
- Plants have also evolved the means to conserve water, notably through the formation of a waxy cuticle and elaborate controls of stomatal apertures, which include the pigment cryptochrome and blue light, and the plant growth substance, abscisic acid.

Plants are cool, thanks to their smooth systems for handling large quantities of water, but their coolness is only one aspect of their sangfroid!

4   Nutrition for the healthy lifestyle

INTRODUCTION

Increasingly, people are thinking about and acting on what they include in their diets. There is a constant bombardment from health specialists, through the media and the internet, exhorting us to maintain a balance in what we eat, an important component of a well regulated lifestyle. Increasingly, we are learning that to eat immoderately and injudiciously is harmful.

We are also generally aware that plants become sick, just as animals do, when not supplied with the nutrients needed for good health. For animals, these requirements are elaborate and include the balanced provision of complex molecules in their diets, such as carbohydrates, fats, and proteins, as well as vitamins and certain minerals. Plants are different in being able to produce their own organic molecules from simpler, inorganic ones.

But, in common with animals, plants need certain minerals for healthy growth. Some of these are essential to both plants and animals in greater or lesser amounts; others are essential either to animals or plants but not both; and some are of probable, but at present uncertain, value to either.

ESSENTIAL MINERAL ELEMENTS

We eat some plant materials for their mineral content, like bananas for their potassium and spinach for its iron, nutrients which are essential both to animals and plants. Sodium is essential to animals but is required for only C_4 plants. Molybdenum is essential to plants but is toxic to animals when more than a trace is present in food. Aluminum, although found in both plants and animals, is not essential to either, is toxic to plants, and may even be detrimental to animals.

MINERAL DEFICIENCY

The symptoms of mineral deficiency in humans are often easy to see and simple to reverse. If we are anemic we take iron for the blood. An incorrect balance of calcium, phosphorus, and vitamin D can lead to rickets in children or osteomalacia in adults. We use iodized salt and eat iodine-rich foods, such as fish from the sea and the seaweed, kelp, to avoid iodine deficiency, which can lead to the condition called goiter in the functioning of the thyroid gland. Herbivorous wild and domesticated animals seek out, or must be provided with, 'salt licks' to satisfy their craving for sodium. These and many other examples point to a recognition of the need animals have for certain minerals.

Similarly with plants, certain abnormal conditions can be recognized as being linked to a mineral deficiency. "Heart-rot" of sugar beet (boron deficiency), "dieback" in citrus trees (copper), "whip-tail" of cauliflower (molybdenum), and "little leaf" of apple (zinc) are all recognized symptoms which can be eliminated by appropriate treatment.

ANCIENT VIEWS OF NUTRITION

Ideas about nutrition can be traced back at least to Aristotle (384 BCE–322 BCE), who believed that plants and animals took in food in various combinations of the four elements: earth, air, fire, and water. Since plants did not have the kinds of organs animals had, stomach and intestine for instance, for changing the food taken in to fit their purposes, Aristotle inferred and deduced that they must absorb it in a form perfectly suited to their needs. To Aristotle, further evidence for this was the fact that plants produced no waste as excrement, unlike animals. Plants simply sucked in perfect "nutrient fluid" from the earth.

The Greek poet, Virgil (70 BCE–19 BCE), in Part 1 of his poem, *Georgics*, recognized, but probably did not understand, the value of adding nutrients back to arable soils:

> ... do not blush
> to lave the parched acres with luxuriant dung
> nor to scatter the grimy ashes over exhausted ground.

These ancient vague or erroneous views of plant nutrition began to erode in the sixteenth and seventeenth centuries as the scientific style of investigation took hold. Until that time, carrying out experiments was believed to be beneath the dignity of the great and powerful of the day. To question the opinions of such revered sages as Aristotle was regarded as heresy.

THE TURNING POINT

Jean-Baptiste Van Helmont

A major turning point was the classic experiment of **Van Helmont** in the seventeenth century (but not published until 1748) in which he concluded, erroneously as it turned out, that the whole substance of a plant is formed from water alone (see Chapter 1 for details). Van Helmont showed that, although his willow tree increased in bulk by nearly 75 kg over 5 years, the soil in which the tree was grown decreased in weight by only a few grams. He concluded that, therefore, the increased mass of the willow could not possibly have come directly from the soil, as Aristotle had proposed; his result raised many questions about how plants obtain their food.

Challenges to Van Helmont – John Woodward

One of the first to challenge Van Helmont's final conclusion about the role of water in plant growth was **John Woodward**, a Professor of Medicine in London at the end of the seventeenth century. Woodward questioned whether the nutrition of plants came from water itself or from what the water contained. When he grew spearmint, potatoes, and vetch in water from various sources (springs, rivers, rain, sewer effluent, and after purification by distillation), Woodward found that plants grew better in water containing impurities than in pure water. In his own words, most of the water that enters a plant:

> passes through the pores of them [*the stomata of leaves*]
> and exhales up into the atmosphere; that a great part of the

terrestrial matter [*solids, including minerals, from the soil*] mixt with the water passes up into the plant along with it; and that the plant is more or less augmented in proportion as the water contains a greater or smaller quantity of that matter.

Woodward concluded that water alone was not sufficient to sustain a plant; the "terrestrial matter" the water had dissolved in it was also important.

Theodore de Saussure

The issue remained thus until the early part of the nineteenth century when a Swiss scientist, **Theodore de Saussure**, became one of the first to attempt a more systematic investigation of plant mineral nutrition. For example, he grew lady's thumb (*Polygonum persicaria*) in water which had in it a single mineral (potassium, iron, etc.) and found that not all were absorbed by the plants in equal amounts. He also discovered that particular minerals among those absorbed were essential to the growth of his plants while others were not.

Carl Sprengel

These careful studies generated a wide debate among scientists in the first half of the nineteenth century and led to much additional evidence in support of de Saussure's views. One of the most significant further conclusions came from a German investigator, **Carl Sprengel**, who wrote that a soil may be favorable in almost all respects:

> yet may often be unproductive because it is deficient in one single element that is necessary as a food for plants.

Here was a clear statement for the first time of the need plants have for a balanced array of nutrients, the lack of any one of which hinders their growth. Here also was a hint of agricultural science, the systematic investigation of plants and their requirements for optimal growth.

Jean-Baptiste Boussingault

Investigations in the mid-nineteenth century reached a high point through the efforts of **Jean-Baptiste Boussingault**, who is generally credited with laying the foundation of agricultural science. He was not content to study just the mineral composition of plants as his predecessors had done. He also stressed the importance of knowing the balance between the amounts of each mineral extracted from soil by crop plants and recognized the need to fertilize to maintain also the balance of nutrients in the soils themselves.

MINERAL THEORY OF FERTILIZERS

Justus von Liebig

Boussingault's ideas about the balanced mineral requirements of plants led **Justus von Liebig** to draw up a list of plant mineral requirements, namely:

> Plants live upon carbonic acid [carbon dioxide], ammonia (or nitric acid), phosphoric acid, silicic acid, lime, magnesia, potash and iron.

In 1843, in support of what had become known as his mineral theory of fertilizers, Liebig went further in saying that:

> the crops in a field diminish or increase in exact proportion to the diminution or increase of the mineral substances conveyed to them in manure.

This advice on the fertilization of crops was widely embraced and closely followed by farmers for many years.

John Bennett Lawes, Joseph Henry Gilbert, and Rothamsted

John Bennett Lawes, a wealthy landowner and keen scientist, was led, in 1842, by his interest in their effect on crop growth, to open the first factory for the manufacture of artificial fertilizers. A year later, he hired **Joseph Henry Gilbert**, a chemist, as his partner. Together, they initiated a series of long-term field experiments at the Lawes' family estate, Rothamsted Manor, in England, some of which continue to this day at the Rothamsted Experimental Station. Their aim was to put agricultural research, especially the study of plant nutrition, on a thoroughly scientific and statistical basis.

As a result of their field experiments between 1843 and 1847, Lawes and Gilbert challenged Liebig's conclusions (above), insisting that:

- mineral fertilizer alone (e.g. manure) was insufficient to increase crop productivity;
- their results showed clearly that additional nitrogen was key to increasing output of grain crops;
- manure contains some nitrogen (mostly as ammonia) but not nearly enough to bring about the crop yields achieved through heavy fertilization with additional nitrogen.

Through these and subsequent experiments Lawes and Gilbert set enduring standards and benchmarks in research at their Rothamsted base.

Liebig's law of the minimum

Despite being eclipsed in this instance by the careful field studies of Lawes and Gilbert, Liebig made many important contributions to agricultural chemistry, it must be emphasized, one of which was to develop from the work of Sprengel (above) and others, what became known as, **Liebig's law of the minimum**, a most important principle:

> That growth is limited not by the total resources available to organisms but by the scarcest resource.

WHICH MINERAL ELEMENTS ARE ESSENTIAL?

Questions about which minerals from soils are essential to the good health of all plants continue to arise even now although we have a much more complete understanding of the matter than did scientists in Liebig's day. One major problem in determining whether a particular mineral is essential or not to plants was greatly aided by the development of **hydroponics**.

Hydroponics – Julius von Sachs

In the latter half of the nineteenth century, **Julius von Sachs**, a German botanist, developed a way to grow plants directly in aqueous solutions of mineral nutrients finally putting to rest the notion that plants required soil as a source of nutrition. Growing plants in liquid culture (hydroponically) has remained a favorite technique in plant nutrition for both legitimate and illicit purposes!

A major advantage of liquid culture is that it eliminates the problem of soil complexity. Soils contain most of the known chemical elements but plants, although picky about how much of a particular mineral they take in from the soil, are not good at excluding any one of them completely. Tiny amounts of all minerals in a soil are likely to find their way into a plant; over 60 of the chemical elements have been detected in plants.

The ash elements

The mineral content of solutions, unlike soil, can be very closely, if not completely, controlled. Starting with pure water, known amounts of particular minerals can either be added or left out; von Sachs did exactly that.

- After determining the combination of minerals that sustained a plant through its life cycle, he then left out each mineral in turn and observed the effects
- In this way he, and those who followed him, discovered that in addition to carbon, hydrogen and oxygen, plants also required the so-called seven **ash elements**: phosphorus, potassium, nitrogen, sulfur, calcium, iron, and magnesium.

Most of the solid bulk of a green plant is composed of just a small number of chemical elements. The "big four" (carbon, hydrogen, oxygen, and nitrogen) make up 95% of a plant's dry weight. When plant material is incinerated at very high temperature, these four are largely burned off (except for some of the nitrogen), leaving behind the ash elements. Well into the twentieth century, these seven were thought to be the only other minerals essential to plants.

Macronutrients and micronutrients

What early investigators in the field of plant nutrition could not know is that some minerals are required by plants in the tiniest imaginable quantities. For instance, it is estimated that a plant requires 60 million times *less* molybdenum than hydrogen, yet both are essential. Such minute amounts could not be detected by the crude methods of measurement at the time the ash elements were discovered.

Now it is clear that, in addition to the seven ash elements, all of which are required in relatively high amounts except for iron, and which are now known as the **macronutrients**, there is another group of minerals, the **micronutrients**, needed in only tiny quantities, which includes molybdenum, copper, zinc, manganese, boron, chlorine, and nickel.

The number of essential elements needed for growth by higher plants currently stands at 17, but whether the list is complete is impossible to determine; others may be found to be essential in even smaller amounts than molybdenum.

There is also a small group of four so-called **beneficial elements** (sodium, silicon, selenium, and cobalt), which are known to be essential to some plants. It may be that, at some future time, one or more of these will be shown to be universally required.

DEFICIENCY SYMPTOMS

Plants respond to an inadequate supply of one of the essential minerals by showing deficiency symptoms (following Liebig's law of the minimum). Deficiencies include such signs as stunted growth of roots, stems or leaves, yellowing of leaves, and browning of various plant parts. A problem is that certain plant diseases can cause similar distortions. Knowing which are a sign of mineral deficiency and which of disease takes practice and experience.

The appearance of deficiency symptoms does, however, help professional agriculturalists, horticulturalists, and foresters decide which fertilizers to add to their crops, and when.

Deficiencies in older tissues: example – magnesium

Older leaves of a plant turn yellow when magnesium is in short supply. Magnesium is a component of the chlorophyll molecule; a shortage of it limits chlorophyll formation. This element is also very mobile within the plant; when it is in short supply, a plant will destroy chlorophyll in its older leaves, which turn yellow, and move the magnesium to younger, more vigorous leaves developing at shoot tips. Of course, if no more magnesium becomes available, through addition of fertilizer, for example, younger leaves eventually turn yellow, too.

The appearance of deficiency symptoms in older parts of a plant before younger ones indicates that the mineral in short supply is mobile within the plant. Plants prefer to move minerals from older to younger tissues if supply from the environment is insufficient for their needs, examples being nitrogen, phosphorus, potassium, chlorine, and magnesium.

Deficiencies in younger tissues: example – calcium

Appearance of deficiency symptoms first in younger plant parts occurs because some minerals, once used, are released and moved elsewhere only with difficulty.

Once calcium has been used it is difficult to release in part because much of it is built into the structure of an organism. In animals like ourselves large amounts of calcium are fixed in bones from which it is hard, but not impossible (think of osteoporosis), to extract. In plants the "glue" used to cement together cells contains calcium. This is the pectin gel boiled out of fruits during the making of jams and jellies. Once this glue has been formed and put in place, the calcium it contains is unlikely to be released, otherwise the whole plant would become "unglued." Thus, new areas of a plant show calcium deficiency symptoms (twisted and deformed stems, roots, and leaves) before older tissues.

Like calcium, a few other elements are also difficult to move around but the remaining essential minerals cannot be categorized quite so easily. Their deficiency symptoms are not identified so directly with either older or younger tissues – rather they can occur in any or all parts of the plant body.

Soil conditions and deficiency – iron chlorosis

Deficiency does not always mean that a mineral is in low concentration in the soil; deficiency symptoms may develop because of conditions in the soil itself. For example, most soils are rich in iron, yet plants growing there may show signs of iron deficiency, the main symptom of which is yellowing (chlorosis – failure to form sufficient chlorophyll) of leaves.

Volcanic soils are especially rich in iron, their reddish–brown color being an indication of this. It might be expected, therefore, that plants grown in soils of volcanic origin would not show iron deficiency symptoms, but they do. In the Hawaiian Islands, for example, the pineapple crop is sprayed with iron solutions several times in a growing season to prevent leaves from becoming chlorotic. The reason – Hawaiian soils also contain high levels of manganese; iron is not taken in by plants very efficiently when manganese is also abundant.

"Limestone chlorosis" is caused by the fact that in alkaline or pH-neutral calcareous soils, ferric ions (Fe^{3+}) form an insoluble hydrous oxide ($Fe_2O_3.3H_2O$), which is unavailable to plants.

Large applications of phosphate fertilizer can cause iron chlorosis even in soils rich in iron; iron and phosphate react together in the soil to form substances that plant roots cannot absorb.

MINING AND RECYCLING OF MINERALS BY PLANTS

Plants play an irreplaceable role in making minerals available to other organisms. The roots of plants invade almost every particle of soil within a large volume, often to considerable depths, in search of water. At the same time the plant also takes in whatever is dissolved in the water, including minerals. After serving an array of functions within the plant, some of the minerals find their way back to surface layers of the soil during autumn or year-round leaf and twig fall. Death of whole plants leads to the wholesale delivery of minerals into the upper layers of the soil.

Minerals are, thus, "mined" from deep in the soil and deposited at the surface where, when plant litter is mulched, the minerals released can be used either by other plants or by other soil organisms. Also, by eating plants, herbivores and microbes satisfy their own mineral needs after which they, too, return material to the soil surface as waste or when they die (see Part V).

Importance and efficiency of recycling

In this function of cycling minerals from deep in the soil to where they are available to all other living organisms we can say that, as in photosynthesis, plants are indispensable to the stabilization and future of life at the Earth's surface. Without constant retrieval of essential minerals from the depths of the soil, leaching by rain would lead eventually to a depletion of nutrients at the surface and the severe curtailment, if not elimination, of life there.

Dependence on soil acidity

The efficiency with which plants carry out "mining" also depends on soil acidity. The rate of growth of roots is favored by mild acidic soil conditions. We also now understand that most roots of deciduous plants are surrounded by and live in close contact with particular fungi (mycorrhizae, root fungi), which greatly aid the roots' ability to take in minerals and, probably, water. The fungi act as a kind of extended root surface that delivers to the root, from a larger zone of soil, minerals that they have taken in but which are surplus to their needs. Growth of these fungi is also favored by acidic conditions.

Finally, soil acidity favors weathering of rocks and the release from rock particles into the soil of minerals like potassium, magnesium, calcium, and manganese, and water. In addition, carbonates, sulfates, and phosphates are more soluble in soil solution under acidic conditions. All these factors contribute to increasing the availability of minerals in the soil root zone (the **rhizosphere**; see Chapter 3 and Part V for more detailed discussion of the points made in this section).

SUMMARY

- Good nutrition involves provision of certain foods and dietary supplements to all organisms
- For animals these requirements are elaborate and overlap with the simpler needs of plants at least in the requirement for certain essential minerals
- Currently, there are known to be 17 macronutrients and micronutrients essential to all plants
- The amounts of minerals taken in are related closely to the needs of the organism. For example, the ratio of the amount of calcium in the **soil** to that found in **plants** and in **humans** is 1:8:40, respectively; for phosphorus, 1:140:200; and for sulphur, 1:30:130
- These ratios confirm what de Saussure discovered long ago; each kind of living thing takes in the minerals that are essential to it in the amounts needed to maintain optimum growth. But, growth may be limited:

  - either by the fact that one essential mineral is in short supply (Liebig's law of the minimum);
  - or because, even though a mineral may be abundant in the plant's environment, it is not present in a form that the plant can absorb
- Minerals are taken up by a plant in various ways, including simple and facilitated diffusion **down** concentration gradients or by chemiosmosis, an energy-driven mechanism dependent on a proton-motive force, which allows the accumulation of mineral elements **against** a concentration gradient (see Box 4)
- Apart from satisfying their own needs, plants also play a crucial role in recycling nutrients from deep in the soil to the surface where the majority of organisms live, thus ensuring access to the minerals needed to sustain the biosphere.

BOX 4. HOW PLANTS TAKE UP MINERALS

*Mineral ions must first be taken up from the soil across the outer (plasma) membranes of root cells before being moved between compartments within cells (vacuoles, mitochondria, chloroplasts, etc.) or to the rest of the plant. Smaller ions (e.g. K^+, Na^+) may penetrate the lipid bilayer of cell membranes by **simple diffusion** but only slowly; membranes also have embedded in them **transport proteins** which assist the rapid transport across them of ions by **facilitated diffusion;** both modes of transport, being diffusional, can only be used to move an ion down a concentration gradient. Accumulation of higher amounts of minerals by plants from lower concentrations in the soil requires energy input.*

*   **Peter Mitchell**, a British scientist and Nobel laureate, proposed the theory of **chemiosmosis** in 1961 to describe how ATP is synthesized by microorganisms, mitochondria in respiration, and chloroplasts in photosynthesis. In all these cases, he proposed, protons (H^+) are pumped across, and accumulated on one side of, a membrane (bacterial, mitochondrial, or thylakoid, respectively). The energy to do this may come from the breakdown of sugars in respiration (bacteria, mitochondria) or from light in photosynthesis (thylakoid); the outcome is the creation of an energy gradient across the membrane in the form of accumulated H^+ on one side of it. In ATP formation, the stored protons are moved back across the membrane through a special enzyme (**F-type H^+-ATPase**) embedded in it; the energy released is used to form ATP.*

*   *Similar enzymes (called **P-type H^+-ATPases**) work in reverse, i.e. they use ATP to **pump** and accumulate H^+ across a membrane. The energy gradient created, the **proton motive force** (pmf), has two components:*

$$pmf = \Delta\psi - 59\ \Delta pH$$

BOX 4. (cont.)

where Δψ is the normal membrane potential of a plant cell and ΔpH is the [H⁺] gradient. A ten-fold difference in [H⁺](one pH unit) at 25°C contributes 59 mV to a membrane potential; the [H⁺] gradient across the plasma membrane in plants is about

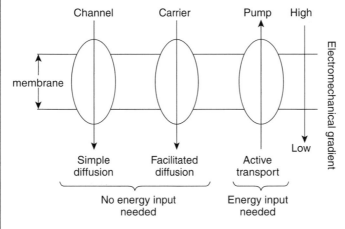

FIGURE 4 Channel, carrier, and pump proteins function to transport substances across cell membranes. The first two are passive processes: transport is either by simple or facilitated diffusion **down** an electrochemical gradient. Pumps use energy to move substances across membranes **against** an electrochemical gradient.

1.5–2 pH units (Figure 4).

Proton gradients across membranes are coupled in all living cells to cellular work of many kinds, including pumping ions from the soil into roots via proteins specific to particular anions (e.g. NO_3^-, SO_4^{2-}) or cations (e.g. K^+, Ca^{2+}, Fe^{3+}) embedded in plasma membranes. The use of H^+ gradients by cells to do work is of apparent ancient origin. The earliest organisms, the **Archea**, *are thought to have harnessed the protons found naturally in water as a limitless, free source of energy for cellular work.*

5 Nitrogen, nitrogen, everywhere ...

INTRODUCTION

Apart from water, nitrogen, among all the essential mineral elements, is *the* key substance limiting where and how well plants grow. The distribution of animals is also linked to nitrogen since animals are dependent on plants for food, directly or indirectly. Why is nitrogen so crucial to the living world?

The machinery used to build, drive, and sustain all living systems is directed from nucleic acid blueprints, the genetic program in DNA and RNA. The machinery itself is made almost exclusively from protein, notably the enzymes which direct the thousands of chemical reactions in living things. Both nucleic acids and proteins contain nitrogen.

The quantity of nitrogen available in usable form is a major determining factor, therefore, in how much nucleic acid and protein organisms can make. Since there is much more protein in an organism than there is nucleic acid, it is the limit to protein production that is the most critical.

SOURCES OF NITROGEN

THE ATMOSPHERE

Nitrogen is enormously abundant. Close to 80% of our atmosphere is made up of nitrogen gas and even that is only about 7% of the total nitrogen on Earth. Nitrogen was given the name "azote," a word meaning "without life," by Antoine Lavoisier, to contrast it with the other major gas in the atmosphere, oxygen. Lavoisier found oxygen to be very active, nitrogen gas, inert. Only when combined with other elements, like hydrogen in ammonia or oxygen in nitrites and nitrates, does nitrogen become more reactive. The three covalent bonds (N≡N) holding the two atoms of nitrogen together in a molecule of the pure element are so strong that very large amounts of energy are needed to break them and release the N atoms so that they can react with other substances.

THE ROCKS OF THE EARTH

In addition to nitrogen as a gas, the enormous quantity of the element trapped in the rocks of the Earth as mineral salts is continually being set free into the soil, waterways, and atmosphere by the slow processes of weathering of rocks by wind and water and by the more explosive but haphazard process of volcanic action, which may blast dust and gases containing nitrogen around the entire globe. Of the total Earth's nitrogen (an estimated 57×10^{18} kg), 93% is locked away in rocks. Most of the rest (about 3.8×10^{18} kg) is in the atmosphere. Only a fraction of 1% (1.5×10^{15} kg) is found in soils and waterways in a form, mainly nitrates, useful to living things.

Thus, the reason that nitrogen is not accessible to most organisms is because it is either locked away in the rocks of the Earth or is in the form of an inert gas in the atmosphere (see also Part V).

Inorganic versus organic sources of nitrogen

Even when nitrogen is available it is still not accessible to many kinds of living things. Animals, including humans, are unable to use inorganic sources of it; nitrogen must be in an organic form before animals can access it. An infant born at, say, 3.5 kg will become an adult human of 60 kg or more; the difference includes about 11 kg of protein. To make this protein we, and other animals, must take in nitrogen either as protein or some other form of organic nitrogen in food.

Protein turnover and renewal

Not only does the average adult human body have in it some 11 kg of protein at all times but this protein is also constantly being broken down and rebuilt. We refer to this as protein "turnover," the constant renewal of the machinery of life, a process common to all living things. During this breakdown and rebuilding some nitrogen is lost from the body. Thus, even a mature adult whose weight remains constant requires a regular supply of dietary protein to replace these losses.

In different tissues, and for different proteins, the renewal process occurs at intervals ranging from minutes to days to months. The turnover of protein in the average adult is roughly 300 g a day, of which about 40 g is lost mainly in urine, feces, sweat, or menstrual fluid, and the shedding of skin, nails, or hair. A diet, therefore, should contain at least 10% protein to replace nitrogen lost. Also, not just any old replacement protein will do. Most importantly, the proteins humans eat must also contain an adequate supply of the *essential* amino acids.

THE *ESSENTIAL* AMINO ACIDS

Proteins are built by joining together amino acids into long chains. Twenty different amino acids are needed to build all the types of proteins found in living systems. Plants can manufacture all 20 for themselves but animals can make only some; the rest must come from food.

Humans are unable to make 10 of the amino acids they need. Gaining these **essential amino acids**[1] is not a problem for those of us with rich diets; the crops and livestock which supply the bulk of our energy, vitamins, and minerals provide our protein as well. In rich countries, and in those specializing in raising livestock, people obtain about one-third of their energy from animal products and nearly three-quarters of their protein.

In poorer countries, animal products typically supply less than one-tenth of daily energy and only about one-eighth of the protein.

---

[1] Lysine, leucine, isoleucine, phenylalanine, tyrosine, methionine, cystine, threonine, tryptophan, and valine.

Poor people often do not have access to adequate supplies of meat and depend on a diet of plants and plant products. Worldwide, especially in poorer countries, plants provide the human race with about 80% of its daily energy and two-thirds of its protein, as well as useful amounts of minerals and vitamins.

CEREALS ARE BEST, EXCEPT ...

Cereals are the best foods since they provide energy in the form of starch and, sometimes, fat, as well as vitamins, and 8–12% protein but they usually lack the essential amino acid, lysine. Some varieties also have less than 10% protein content. This may not create a critical problem for adults although mature men and women in underfed regions of the world undoubtedly benefit from extra protein. The main problem is children, who need proportionately more protein than adults. Underfed children suffer from the disease kwashiorkor, a direct result of protein deficiency.

But, at least in the case of lysine deficiency, a remedy is to hand.

PULSE CROPS

In all parts of the world, especially where plants are the main source of food, crops such as the "pulses" are cultivated to supplement the diet. Beans, peas, lentils, and chickpeas are pulse (legume) crops of particular value, the protein of which complements that of cereals; their seeds are rich in lysine.

As far back as early biblical times, King Nebuchadnezzar ordered Daniel and the other Israelites to eat the King's meat and drink the King's wine, but Daniel refused, saying:

> let them [the Babylonians] give us pulse, and water to drink (KJV).

Nebuchadnezzar relented when Daniel went on to prove the superiority of his own diet over that of the King, who then ordered his servants to:

[take away from the Israelites] the portion of the king's meat, and the wine that they should drink; and [give] them pulse (KJV).

Pulses and cereals together provide an excellent protein balance. All the world's basic cuisines include mixed cereal–pulse dishes: for example, chapatis and dhal in India; frijoles and tortillas in Mexico; even beans on toast. Soybeans have become one of the most important of all cash crops because they can be processed in so many ways. Textured soy protein can be flavored to make anything from dog food to hamburgers. Oil and "milk" are also extracted from the bean and used in many products. In Japan, miso, a breakfast soup, is made from fermented soybean paste; tofu, a soybean milk product, is also used in soups; soy sauce is just fermented beans.

PLANTS AND THE QUEST FOR NITROGEN

Carnivorous plants solve the problem of nitrogen acquisition in a straightforward way: they turn the tables on animals, eating them instead of being eaten.

Insects attracted to pitcher plants by the sweet smell of their nectar slide helplessly down the slippery surface of a bell-shaped funnel into an acid bath at the base where they drown. As the flesh of the animal decomposes, the nitrogen released is absorbed by the plant.

Pitcher plants and other carnivorous relatives, such as the Venus flytrap, sundews, and bladderworts, live in bogs where the ground is so acidic and waterlogged that bacteria cannot completely break down dead plant matter into nitrogen-rich soil. A far less nutritious peat is formed which cannot supply the plant inhabitants of the bog with the nitrogen they need. As a result, carnivorous plants draw their nitrogen from animal prey.

All have their individual ways of attracting and, then, engulfing their victims. But the goal remains the same in all cases: the quest for nitrogen.

THE NITROGEN CYCLE

For all other plants, understanding how they gain nitrogen requires that we first know something about the "nitrogen cycle."

In brief (see Part V for a fuller account), the nitrogen cycle starts in the soil where dead plants and animals are decomposed by microbes, primarily bacteria and fungi, releasing nitrogen in the form of nitrate and ammonia. Some of the nitrogen is returned directly to living plants via their roots, but not all. Some is converted by soil microbes to nitrogen gases, which drifts off into the inert nitrogen pool of the air and are lost to living systems. Globally, an estimated 93–190 million tonnes of nitrogen are lost to the atmosphere in this way annually.

Obviously, if this were to continue, all the nitrogen available to living systems would be lost to the atmosphere eventually. That this does not happen is because some organisms have found a way to take nitrogen directly from the air, breaking the very strong covalent bonds between the nitrogen atoms and then converting the nitrogen to a form usable by living things. This process of **nitrogen fixation** will be discussed below but, first, the release of nitrogen into the soil.

Soil nitrogen – mineralization and nitrate

The nitrogen released by decomposition in soil is converted to ammonia (NH_3) by soil microbes. In the case of animal flesh this process, called **putrefaction**, produces not just ammonia but also other, yet more foul-smelling, products. The breakdown of plant material is less obnoxious, may take several years to accomplish, and gives rise to **humus**.

In most soils, the ammonia produced does not stay around for very long, which is fortunate since it is toxic to both animals and plants. Nitrifying bacteria in soils quickly convert ammonia to nitrite (NO_2^-), then nitrate (NO_3^-), extracting energy from the ammonia for their own needs, a process called **mineralization**.

The advantage of nitrate to a plant is that it is not toxic; the disadvantage, that it is soluble in water and quickly leached from soil by rain or irrigation. Then, it either percolates deep into the soil where it can no longer be reached by plant roots or is carried in waterways to the ocean. In agricultural practice, this lost nitrate is replaced either by adding fertilizer or through crop rotation with legumes which, as we shall see shortly, return nitrogen to depleted soils.

Under some conditions, soil bacteria can cause the release of nitrogen from nitrates in the form of nitrogen gas. Waterlogged soils, for example, become short of oxygen. Some soil bacteria in flooded fields and pastures respond to lack of oxygen for respiration by breaking down nitrate, seeking the oxygen in the nitrate, not the nitrogen. They take the oxygen for themselves, releasing nitrogen gas as a byproduct into the atmosphere.

Eutrophication

Overuse of fertilizers has led to leaching of excess nitrate and phosphate into lakes, rivers, and streams in many parts of the world. In waterways, high concentrations of nitrate and phosphate can cause **eutrophication**, an explosive growth of algae and aquatic plants. When these masses of vegetation die they are decomposed by equally excessive numbers of microbes. The microbes use up much of the dissolved oxygen in respiration, leaving an insufficient amount for fish and other aquatic animals, which then also die, often quite suddenly and in great numbers, leaving the water devoid of life.

The "killing" of natural waterways by eutrophication is a great threat to the environment anywhere that profligate use of fertilizers occurs, leading in the oceans, for example, to **dead zones** (see also Chapter 16). Nitrogen may be essential to plant life, but an excess of it can be highly detrimental.

Nitrate as a hazard

Nitrate in moderate quantity may be harmless to plants but not to infants. Nitrate leached from soil into ground water or waterways may end up in drinking water. Bacteria in our intestines can convert this nitrate to nitrite which binds to hemoglobin. The strongest bonds are formed with fetal blood, which persists in infants for some time after birth. Hemoglobin with nitrite bound to it no longer functions leading to anemia in very young children.

REPLENISHING SOIL NITROGEN
Either by leaching of nitrates from soil or by the breakdown of nitrates to form nitrogen gas, there is a continual drain of usable forms of nitrogen away from the biosphere. Fortunately, ways exist to replenish soil nitrogen, both artificially (the Haber–Bosch process) and naturally (nitrogen fixation).

THE HABER–BOSCH PROCESS
A major impetus leading to the setting up of the first industrial plants to produce ammonia from nitrogen gas was the need by Germany for munitions in World War I.

- Hydrogen must be added to nitrogen to convert it to ammonia. In 1909, **Fritz Haber**, a German chemist, found a way to do this efficiently
- **Carl Bosch**, a German industrial chemist, set up the first commercial plant to manufacture ammonia using the Haber process in 1913
- At high temperature (300–400 °C) and pressure (35 MPa), in the presence of a catalyst, nitrogen gas and hydrogen can be made to combine to form ammonia
- Natural gas, coal gas, or petroleum are readily available sources of hydrogen
- Today, about 80 million tonnes of ammonia are produced by the Haber–Bosch process for peaceful use in fertilizers; the main product, anhydrous ammonia, can be applied directly to soil or converted first to products like nitrate or urea.

NITROGEN FIXATION
As explained earlier in this chapter, the mineralization of animal and plant material in the soil leads to the recycling of nitrogen in a form that can be reabsorbed by living plants. These plants may be eaten, in turn, by animals to satisfy their own nitrogen needs. But this recycling process does not lead to the addition of *extra* nitrogen to

the biosphere. The weathering of rocks adds some, slowly. Lightning, in tens of thousands of electrical storms around the world, as well as ultraviolet radiation, creates nitrogen oxides in the air which can be washed down to earth dissolved in rain, providing about 10% of the nitrogen fixed annually. In addition, these days, air pollution containing nitrogen oxides from automobile and industrial emissions can also be washed down as acid rain (nitric acid), adding yet more nitrogen to soil.

But these sources are minor on a global scale. Fortunately, nature has devised an alternative upon which the natural world depends and which is hugely exploited in agriculture as well.

The value of legumes

Earlier, it was pointed out that the seeds of the pulses can be used directly in diets to supplement the protein provided by cereals. Legumes in general, both wild and cultivated, including the pulses, can also add nitrogen to soil both as the plants grow and when they die and decay. The beneficial effects on the soil of legumes and the importance of green manuring were realized by the ancient Chinese, Greeks, and Romans. For example, the Roman poet, Virgil (70–19 BCE), recommended rotation with legumes in Part 1 of *Georgics*:

> ... sow golden spelt [an ancient grain crop]
> where late you gleaned the bright bean in its trembling
> pod, the blooms of slender vetch, or the bitter
> lupine with frail stem and whispering leaves.

The use of legumes in crop rotation was well established, therefore, long before the reason for their beneficial effect was understood.

Jean-Baptiste Boussingault

Jean-Baptiste Boussingault (1802–1887) provided the first firm evidence for the fixation by legumes of nitrogen directly from the air.

- He grew crops like clover and peas in soil which had no available nitrogen, yet the plants somehow gained an ample supply of nitrogen for their needs
- When he grew cereals like wheat and oats in the same type of soil they gained no nitrogen
- Boussingault concluded that the nitrogen gained by the legumes "is derived from the atmosphere; but I do not pretend to say in what precise manner the assimilation takes place."

For several decades after Boussingault's discovery, as work on the role of nitrogen from the atmosphere in plant nutrition continued, increasingly scientists became convinced that legumes could "fix" nitrogen from the air.

The two Hermanns: Hellriegel and Wilfarth

The problem remained baffling, however, until the late nineteenth century when two German investigators, **Hermann Hellriegel** and **Hermann Wilfarth**, discovered the role of bacteria in the nodules so characteristic of legume roots.

- When Hellriegel and Wilfarth grew peas in soil sterilized by heating they were found not to grow well and had no nodules on their roots
- Others grown in similar sterile soil to which was added an aqueous solution, a leachate, taken from unsterilized soil in which peas had been grown, grew well and developed root nodules
- Nodules examined microscopically were teeming with bacteria, which led Hellriegel and Wilfarth to conclude that the microbes somehow took nitrogen from the air and "fixed" it into a form useful to the plant. How this was achieved remained a mystery for many more years.

Between 150 and 190 million tonnes of nitrogen fixed globally each year is accounted for by the reduction of N_2 to NH_3 by living organisms. Two major groups of microbes can fix nitrogen from the air.

Free-living N-fixing bacteria

First, there are the microbes that live free in the soil or in waterways, among which are the **cyanobacteria**, found on wet surfaces, including soil. Other cyanobacteria occur in loose association with a variety of plant species, such as one that lives inside the leaves of a tiny fern, *Azolla filiculoides*, which floats on the water in rice paddies. The fern, with its associated cyanobacterium *Anabaena*, provides most of the nitrogen needed by the rice crop. Yet other cyanobacteria live on or in the roots of several evergreen plants as well as inside a number of ferns, lichens, and liverworts. The cycad, *Macrozamia*, found in eucalyptus woodlands in Australia, fixes nitrogen through an association with cyanobacteria lodged in its above-ground roots.

Certain loose associations between free-living soil microbes (other than cyanobacteria) and plant roots occur around the globe. N-fixing bacteria grow on or near the root surfaces of many crop plants such as maize, wheat, sorghum, rice, sugarcane, millet, and other grasses. The microbes gain by having access to a supply of sugars for their energy needs produced by the plants in photosynthesis. In return, the plants have a guaranteed source of a scarce resource, nitrogen.

Nodular N-fixing bacteria: the rhizobia

The second group of N-fixing microbes are those with a special, intimate relationship with a host plant through the formation of nodules within which **nitrogen fixation** takes place. The Fabaceae (previously the Leguminosae), such as fodder legumes like clover, lucerne (alfalfa), and the vetches, as well as the big-seeded pulse crops such as peas, beans, lentils, chickpeas and groundnuts, fall into this group. These, and shrubs such as gorse and broom, and trees, including the extremely important acacias of Africa and Australia, all harbor bacteria of the **Rhizobium** type in their roots. Rhizobia were, indeed, the microbes in the leachates used by Hellriegel and Willfarth in their early investigations of nitrogen fixation in the root nodules of peas.

Rhizobial bacteria (of which there are many species) live free in the soil. However, unlike other soil microbes capable of fixing nitrogen directly, rhizobia do not do so in their free-living state. They take in nitrogen in the same way as any other soil organism, as nitrate, and will live quite happily in this way until a legume appears close by.

Root nodules – form and function

As legume seedlings develop, their roots secrete substances into the soil which attract rhizobia. The bacteria enter the roots, where they stimulate the formation and growth of nodules, inside which the microbes multiply. At the same time, the bacteria take on different shapes to such a degree that they no longer resemble the soil rhizobia from which they came. These **bacteroids** can also now fix nitrogen, using the enzyme, **nitrogenase** (see Box 5).

BOX 5. NITROGENASE

*Prokaryotes are the only organisms which have the enzyme nitrogenase, required for nitrogen fixation. The first cell-free extracts of the enzyme were obtained in 1960 from the anaerobic bacterium, **Clostridium pasteurianum**, by researchers at the chemical company, Dupont. This achievement allowed, henceforth, increasingly detailed investigation of how the nitrogenase reaction proceeded; the enzyme has now been isolated and purified from virtually all nitrogen-fixing prokaryotes.*

The overall reaction catalyzed by nitrogenase is:

$$8H^+ + 8e^- + N_2 + 16\,ATP \rightarrow 2NH_3 + H_2 + 16\,ADP + 16\,Pi$$

*The enzyme comprises two proteins of quite different sizes: the smaller **Fe protein** contains four atoms of iron and can accept electrons from **ferredoxin**, an important electron carrier both in nitrogen fixation and in photosynthesis. The larger **Fe–Mo protein** contains two atoms of molybdenum and a variable number of iron atoms depending on the species and physiological state of the host prokaryote. It can accept electrons from the Fe protein, in the process reducing N_2 to NH_3 and H^+ to H_2 in a 1:1 ratio. The*

BOX 5. (cont.)

role of ATP, while not entirely clear, seems to be as a source of energy to facilitate the transfer of electrons between the Fe and Fe–Mo proteins.

Fixation of nitrogen is a considerable drain on energy resources. For example, the number of ATP molecules plus the reducing power represented by ferredoxin needed to fix one molecule of nitrogen to ammonia represents about three times the amount of energy required for fixing a molecule of carbon dioxide in photosynthesis. Put another way: it has been found that about 12 g of carbon are required to fix 1 gram of nitrogen in the nodules of soybean. The source of this carbon is carbohydrate formed in photosynthesis and diverted from the plant to the bacteroids in root nodules, a considerable drain on the plant's energy resources.

*Another drain on energy is the inescapable production of hydrogen during the reaction. Between 30 and 60% of the ATP and electrons supplied to nitrogenase may be diverted to hydrogen production, a significant waste of energy which could have been used to reduce more nitrogen. Not all the hydrogen produced is released to the atmosphere, however. Some nitrogen-fixers also have a **hydrogenase**, which couples H_2 oxidation to ATP formation. Currently, using biotechnology, there is interest in improving the capacity of rhizobia to recycle hydrogen and increase the overall energy efficiency of nitrogen fixation in key crop plants.*

The bacteroids accumulate at the center of each nodule, surrounded by tissue that often takes on a pink color caused by the production by the plant of a form of hemoglobin which closely resembles ours but which, because it is found in legumes, is called **leghemoglobin**.

Leghemoglobin

Legumes need a substance like leghemoglobin because N-fixation stops when bacteroids are exposed to the oxygen concentration found in air. On the other hand, nodule bacteroids have as much need for oxygen as any other living thing for respiration. Leghemoglobin acts to control the flow of oxygen so that enough is supplied to nodules to maintain bacteroid respiration without destroying their ability to fix nitrogen.

In our blood, hemoglobin carries oxygen to all tissues of the body and then returns to the lungs for more. Leghemoglobin does not circulate but is contained in the nodule tissue surrounding the area where bacteroids are lodged. As air percolates into nodules from the soil the leghemoglobin intercepts the oxygen but allows nitrogen to pass through to the bacteroid colonies. The leghemoglobin delivers the oxygen it has intercepted to the bacteroids at a controlled rate; just fast enough to satisfy the respiratory needs of the microbes, but not in such a flood as to poison nitrogen fixation.

NITROGEN FIXATION AND THE PHOTOSYNTHESIS CONNECTION

Bacteroids also need a supply of energy to make the nitrogen gas reactive. For this they rely on the host plant, which transports sugars formed in photosynthesis from its leaves to its root nodules. Bacteroids use this source of energy to satisfy their own energy needs and, in return, produce more ammonia than they need. The host plant converts the ammonia to a non-toxic, organic nitrogen form (see below) in the nodule before transporting it to other tissues. Thus, conditions favoring rapid photosynthesis in a host legume plant, such as good moisture, moderate to high temperatures, and bright sunlight, also favor nitrogen fixation.

The rate of nitrogen fixation is at its maximum in early afternoon when the movement of sugar from photosynthesizing leaves to other parts of the plant is most rapid. Early afternoon is also the time of day when transpiration is often at its fastest. The rapid removal of products of nitrogen fixation from the roots in the fast-moving water stream up to transpiring leaves helps stimulate fixation of more nitrogen by root nodules.

Beyond ammonia

The first product of nitrogen fixation is ammonia which, as already noted, is toxic to plants. For this reason, it is quickly converted in roots to **amino acids** and the amide, **glutamine**. The latter is, then, converted to either **asparagine** or the **ureides**, **allantoin** and **allantoic acid**, before being transported to the rest of the plant.

Nitrate taken in from the soil, on the other hand, is not toxic and, therefore, may or may not be converted to amino acids in roots before being moved elsewhere in the plant. Nitrate itself may be transported in the water stream all the way to the leaves before being converted to ammonia, then other nitrogenous compounds.

Nitrogen recycling

Measurements on grasses and legumes indicate that there is extensive daily recycling of nitrogen from roots to leaves and back again. This constant movement is important in directing a balanced supply of nitrogen to all areas of the plant. These movements include not just the redistribution of new nitrogen coming in as nitrate through roots or as ammonia from root nodules but also the turnover of protein nitrogen in all parts of the plant. As mentioned earlier in the chapter, protein molecules in all living systems are constantly being broken down into their amino acid units and rebuilt. Products of turnover can be moved anywhere in the plant that protein molecules are being formed.

Amino acid release and redistribution is especially important at certain times of year. Nitrogen gained by the plant during the leafy growth phase of its life cycle can be mobilized and moved from roots, stems, and leaves into flowers, fruits, seeds, and storage organs as the need arises. Nitrogen is moved wholesale out of dying leaves in the autumn in temperate climates and stored in the roots and stems of the plant for use the following spring. In some annuals, the movement of nitrogen out of leaves can be extensive: wheat leaves, for example, lose up to 85% percent of their nitrogen to flowers and seeds before they die.

THE ENIGMA OF NITROGEN FIXATION

One of the enduring enigmas in biology is why the hugely important ability to fix nitrogen from the air possessed by certain microbes has not been transferred to plants in the course of millions of years of evolution.

One possible answer to this question is that considerably more energy is required for every molecule of nitrogen fixed than is used in fixing a molecule of carbon dioxide in photosynthesis. Thus, most plants do not associate with microbes to fix nitrogen the way legumes do. The great majority of green plants take their nitrogen directly from the soil, mainly as nitrate. Even legumes and other plants which harbor organisms that can fix nitrogen stop doing so when supplied with adequate amounts of nitrate. One of the features of legume crops is that their productivity is not significantly increased by providing them with nitrogen fertilizer because the plant shuts down fixation until the added nitrogen is used up. It is an either/or proposition, not an additive one.

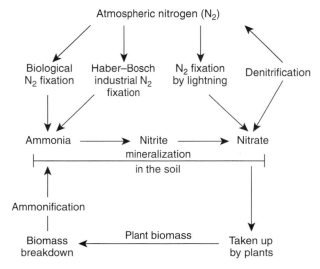

FIGURE 5 The major sources of nitrogen, natural and manufactured, available to plants, and the relationship between them (after Hopkins and Hüner, 2009).

Not that the conversion of nitrate to ammonia has a lower energy requirement than nitrogen fixation. It may simply be that the processing of nitrate occurs within the plant itself whereas nitrogen fixation occurs in association with microbes. Maybe nitrate processing is, therefore, easier to accomplish. There is as yet no fully satisfactory answer to this question (Figure 5).

SUMMARY

- Nitrogen is essential for all organisms. It is used to form such important molecules as DNA, RNA, and enzyme proteins
- It is abundant on our planet's rocks and atmosphere but is present largely in inert forms, which cannot be accessed by living organisms
- Nitrogen is made available to plants, and hence to animals, through the agency of microbes in soil and water bodies
- Free-living soil and waterborne microbes "fix" nitrogen from the air and provide it to other organisms
- Legume plants form closer symbiotic relationships with the rhizobia in nodules on roots within which these microbes also "fix" the nitrogen in air to the benefit of their hosts
- Nitrogen fixation by microbes, through the agency of the enzyme nitrogenase, is, thus, essential to the cycling of nitrogen out of the atmosphere into the biosphere
- Green plants provide a crucial link in this cycling through their direct participation in host–microbe associations as well as by absorbing forms of nitrogen from the environment and converting them to the kinds of organic forms which animals and other non-green organisms can use.

Nitrogen may be everywhere in great abundance, but without the intervention of green organisms it would remain "azote."

6   Transport of delights

DEMANDS ON A PLANT TRANSPORT SYSTEM

Delivery to all tissues ...

Growth can take place in any area of a plant throughout its lifetime (see Part III). However, only green parts of the plant can photosynthesize yet non-green parts also require food to provide energy and other needs. For example, buds at the tips of branches contain many young leaves which are too immature to carry out photosynthesis yet are growing rapidly. The trunks and stems of trees and many shrubs are not green yet contain substantial amounts of living tissues which must be provided with nourishment. Energy is needed to form flowers, fruits, and seeds, none of which may be green, at least when mature. Some plants form food storage organs like bulbs, corms, and tubers, to which, as they swell, food must be delivered.

... At different times

These various demands for food arise not only at widely separated locations in a plant but also at different times during a growing season. In temperate climates early in spring the main need may be to move food from mature leaves capable of photosynthesis or from storage organs to young, developing leaves at stem tips, and to rapidly growing roots. Later, as the plant begins reproductive growth, flowers must be nurtured, followed by fruits and seeds. Later still, if overwintering storage organs are formed they, too, must be supplied with significant amounts of food. All of these organs can be at different locations on the plant body.

Whatever supply system the plant has must be sufficiently versatile to deliver the food required not just downward to roots and storage organs but also upwards to developing stem tips, and across to developing flowers, fruits, and seeds.

... In different directions

Then there are the leaves themselves which, when young, import food to fuel their rapid expansion but which, as they mature and start full production of sugars in photosynthesis, become food exporters. The delivery system must be capable of transporting food **into** the leaf at one time and **out of** it some time later.

DISCOVERY OF CIRCULATORY SYSTEMS – TRANSLOCATION

The discovery of human blood circulation by **William Harvey** in the first half of the seventeenth century stimulated the study of circulatory systems in general, including plants. Yet, only in later decades of the twentieth century was a reasonably complete understanding gained of how foods and other substances are moved around a plant, a process called **translocation**.

The problem, in part, was because translocation takes place in a system of tubes of microscopic size located, mostly, deep inside a plant. In animals, some arteries and veins are visible even to the naked eye but not in plants. The hearts of many animals and associated plumbing are relatively easy to find and study. Even here, a major stumbling block at first to a deeper understanding of how the system worked was the microscopic capillaries in the tissues of the body that connect arteries with veins. Only when it was realized that capillaries are the bridge between the two could the entire circulatory system be visualized.

In plants, there is really *only* a capillary system, which was difficult to find and even more difficult to study closely. There is nothing equivalent to animal arteries and veins in plants although we use the word "vein" loosely to describe the circulatory system in a leaf.

Marcello Malpighi

The Italian scientist, **Marcello Malpighi**, beginning in 1675, made telling observations about the translocation and role of sap in plants.

- He girdled a tree by removing a complete ring of bark from around its trunk without damaging the underlying woody tissue
- Over the following weeks he noticed the bark above the girdle becoming swollen; immediately below, thinner
- Malphigi also showed that ascending, or **raw, sap**, as he called it, from the soil was changed in the leaves, but only in sunlight, into an **elaborated sap**, which was then moved to other parts of the plant
- He found that girdling blocked the movement of the latter, causing the bark above the girdle to swell
- He also noted that swelling was not as prominent in winter when girdled trees were leafless
- Malpighi also discovered that, although the downward movement of elaborated sap was blocked by girdling, water movement from roots to leaves continued. Unfortunately he, and others who followed, did not appreciate the significance of this discovery; they failed to see that the movement of water occurred in different tissues from those used to move elaborated sap
- Well into the eighteenth century it was still believed that ascending and descending streams of fluids in plants traveled in the same channels. Even Stephen Hales, who provided so much of the early information on water movement in plants (see Chapter 3), declared, in his book *Vegetable Staticks*, half a century after Malpighi and after many girdling experiments of his own, that:

> Upon the whole, I think we have, from these experiments and observations, sufficient ground to believe, that there is no circulation of sap in vegetables; notwithstanding many ingenious persons have been induced to think there was, from several curious observations and experiments, which evidently prove, that sap does in some measure recede from the top towards the lower parts of plants, whence they were with good probability of reason induced to think that the sap circulated.

Despite these misunderstandings, by the end of the eighteenth century, it was generally understood that sap from roots was carried upwards to leaves in xylem vessels and tracheids in the wood. There, in the presence of light and air, photosynthesis occurred and the sap was "elaborated," made rich in sugars, then transported to other parts of the plant, providing nourishment for growth. What was not appreciated was how it got there.

PHLOEM AND ITS SIEVE TUBES

During the third decade of the nineteenth century, a new conducting tube system was discovered in the bark of plants that was quite different and distinct from that for transporting water. The tubes used for water transport in wood are simple, open channels much like the pipes which carry water to and within our homes; they are dead tissue. In contrast, the tubes discovered in bark were living cells and, it was soon realized, were important in the translocation of Malpighi's elaborated sap. In bright light the sap in these tubes was rich in sugars; in poor light or complete darkness, sugar content of sap was much reduced.

We now know that these channels, **sieve tubes**, occur in most green plants in the tissue, **phloem**. We know, too, that wherever they are found, sieve tubes perform the same function of carrying predominantly the products of photosynthesis, and some other substances, from places where they are formed or stored, to where they are needed. Organs like leaves, from which food is exported, have come to be known as **sources** of food; **sinks** are those areas of the plant to which food is supplied.

Questions about sieve tubes

Once the significance of sieve tubes was understood, attention could be directed to answering questions such as how quickly the movement of sap can occur in them. After all, if they are living tissue full of living matter, how can they act as pipes for the efficient movement of food around a plant? As already pointed out, we know that food has to be moved in all directions in a plant and that several source–sink channels may have to operate at the same time. How, then, is it possible to have sap moving in sieve tubes in opposite directions, simultaneously?

Finding answers to questions such as these have occupied plant biologists for decades. Understanding how sieve tubes work has very practical value since we now know that not just foods can be moved in them. For example, there is much interest in how to apply to plants chemicals, like herbicides, or substances that promote growth, or fertilizers, directly onto leaves rather than through the soil. Many of these substances must then be transported from where they are applied to other locations in the plant where they have their effect. Sieve tubes play a role in this.

The spread within a plant of many of the diseases caused by fungi, bacteria, and viruses often involves the circulatory system. For instance, certain fungi that infect roots have their effect by producing toxins which are spread up the stem to the leaves via the water stream. Some viruses are introduced directly into the sieve tubes in the leaves or within stems by feeding insects (aphids, leafhoppers, white flies) and then moved along with translocating food.

The main challenge in answering questions about the performance of sieve tubes was how to observe them in a working, functioning state when they are microscopic in size, often deeply hidden inside the plant, and surrounded by other kinds of tissues not involved in translocation.

Sieve tubes are delicate

From the early days of investigation it was learned that sieve tubes are easily damaged if disturbed and stop functioning immediately. This is frustrating for the investigator but makes sense in the functioning of the plant. Damage to its translocation system may be a signal to the plant that it is under attack by disease organisms or predators. The immediate response of the plant to perceived danger is to arrest flow in the damaged section of the translocation system, perhaps in an attempt to prevent the introduction of infection or to protect the rest of the plant from further injury. Whatever the reason, the delicacy of sieve tubes means that they cannot easily be removed from plants to see how they work. They have to be examined *in situ*. Ingenious ways had to be found to gain access to the translocation system without causing it to cease functioning, one of which, the aphid stylet method, is outlined in Box 6.

BOX 6. PRESSURE–FLOW MODEL OF PHLOEM TRANSLOCATION

The pressure–flow model, first put forward between 1926 and 1930 by **Ernst Münch***, demonstrates how the flow of translocate in sieve elements can be as a result of a pressure gradient between a source and a sink ($\Delta\psi_p$). The gradient is created by phloem loading at a source and unloading at a sink. At a source (e.g. a leaf), loading of, primarily, sucrose from photosynthesis into sieve elements in leaf veins leads to a higher solute concentration (a more negative solute potential: $\Delta\psi_s$). This, in turn, causes a sharp drop in water potential ($\Delta\psi_w$), in response to which water is attracted into the sieve elements from adjacent xylem, causing an increase in turgor pressure ($\Delta\psi_p$). Increased turgor pressure forces the mass or bulk flow of translocate along sieve tubes towards a sink (e.g. a non-green part of a plant) where removal and use of solutes lowers turgor pressure, allowing solvent (water) to flow back to the xylem. The relationship between the three potentials involved can be expressed as:*

$$\psi_p = \psi_w - \psi_s$$

Although challenged and modified many times since its introduction by Münch, this model remains the most popular explanation of how phloem translocation occurs, and continues to gain support from a growing number of experimental results.

Crucially, it must be demonstrated that the pressure gradient along sieve tubes is large enough to explain observed rates of mass/bulk flow in phloem. The most accurate measurements of turgor

---

BOX 6. (cont.)

pressure in sieve elements have been made using the aphid sty-
let method devised by two insect physiologists, **J. S. Kennedy** *and*
***T. E. Mittler**, in 1953. Aphids insert their mouthparts (stylet) into*
a single sieve element without disturbing it. Kennedy and Mittler
found a way to anesthetize a feeding aphid with a stream of CO_2,
then, remove its body with a razor blade, leaving the, now, open-
ended stylet still embedded in the sieve element, the contents of
which continue to flow, under pressure, through the stylet.

Measurements using micromanometers attached to exud-
ing aphid stylets show that turgor pressures in angiosperm sieve
elements are typically between 0.1 MPa and 2.5 MPa depend-
ing on species, physiological status of the plant, and time of
day. From measurements of solute concentrations in stylet
exudates, it has been shown that pressure gradients between
sources and sinks are sufficiently steep to account for observed
speeds of phloem translocation over the lengths of pathways
known to exist (e.g. leaf to root).

Observations like these provide strong support for the
pressure–flow model of phloem translocation.

Rate of phloem transport

Using a variety of methods it has been learned that translocation
occurs at a speed of between 50 and 150 cm h^{-1}. By way of analogy, a speed of 100 cm h^{-1} is equivalent to the motion of the tip of a
16-cm-long minute hand on a clock. With patience we can see a rate
of movement like this with the naked eye. Put another way, a sugar
molecule translocated 100 m down a tree from leaf to root at a velocity of 100 cm h^{-1} would take 100 h, a little over 4 days, to reach its
destination.

It is probably not necessary for a plant to move food over quite
such great distances. The lowest leaves are closer to the roots than
those at the top of a tree and are more likely to be the sources from

which the roots are supplied with food. Still, translocation occurs in tubes that are much narrower than the finest strand of hair, which are filled with living material rather than being open pipes, and within which movement of fluids is remarkably rapid.

Pressure-driven system

The rapidity of movement is due to the maintenance of a high pressure in sieve tubes. The high concentration of sugars in the tubes acts strongly to attract water by osmosis (see Chapter 3), maintaining a pressure there of between 2.0 and 3.0 MPa. The pressure is highest in the leaves where sugars produced in photosynthesis are "loaded" into the sieve tubes in leaf veins. The sugar solution is then forced out of the source leaf towards various sinks. As sugar is withdrawn at sinks, the pressure in the nearby sieve tubes drops, creating a pressure gradient from source to sink.

Contents of sieve tubes

Carbohydrates make up at least 90% of the material translocated in sieve tubes, of which sugars, most often sucrose, are the most frequent. Concentrations as high as 30% (w/v; a thin syrup) have been recorded in leaf veins just loaded with sugar following a period of rapid photosynthesis.

This is not to say that other, less abundant, compounds in sieve tubes are unimportant. Some minerals, including nitrogen, usually accompany the sugars and can be just as important for growth activities at a sink. Phloem sap is a balanced mixture, providing good nutrition to non-green parts of plants.

ALLOCATION AND PARTITIONING OF PHOTOSYNTHATE

In general, the flow of food in sieve tubes changes direction simply in response to the location of the sinks, whether they be further up or down the plant from a particular source. But this is not always the case. It has long been known, for example, that leaves lower down a plant tend to transport relatively more food downwards to roots;

upper leaves tend to translocate preferentially upward to the young-est branches, developing fruits and seeds.

How does a plant allocate and partition its food among all competing sinks over its lifetime? This is a question of great interest and practical importance. Somehow, delivery of adequate volumes of food must be maintained to all competing sinks to sustain bal-anced growth of the entire plant body. Crop yields depend on how much food produced in leaves is translocated to and then stored in those tissues of the plant which will be harvested, whether seeds, fruits, or storage organs. Understanding what controls the allocation and partitioning of food within plants is important to breeders of new crop varieties. If the allocation and partitioning process could be understood in detail, then maybe it could also be manipulated to produce higher yields.

Sucrose versus starch

What is understood in more detail now is how plants regulate sucrose and starch production during photosynthesis. **Sucrose** is formed exclusively in the cytosol of a leaf mesophyll cell; **starch** in the chloroplast. When a plant slows its sucrose production, starch accumulates in the chloroplast, and vice versa. The slowdown in sucrose formation was thought to be caused by a fall in demand for the sugar in sinks, but this is not so.

A special form of fructose, **fructose-2,6-bisphosphate** (F-2,6-BP), is used as a signal in leaves to regulate partitioning of carbon assimi-lated in photosynthesis between sucrose and starch. As more sucrose is formed in the cytosol of a leaf, more F-2,6-BP is also produced. As F-2,6-BP concentration increases, sucrose production slows, lead-ing to less photosynthate being released from the chloroplast. The accumulation of photosynthate in the chloroplast leads to the for-mation of more starch there as well as to a moderation in the rate of photosynthesis.

Control of the use of the products of photosynthesis beyond this initial level of their formation is less clearly understood.

Allocation, storage, and use of photosynthate
Mobile photosynthate may be used immediately in metabolism by
the leaf itself. Another portion may be stored for future use in the
leaf either as sucrose or starch. After all, the leaf requires energy and
carbon products for continued growth at night when photosynthesis
is not taking place. Finally, some of it may be exported to sinks. How
the allocation of mobile photosynthate among these competing endpoints is regulated is poorly understood.

In many plants, about half of leaf assimilate is exported to sinks.
During daylight hours, the source may be sucrose formed directly
in photosynthesis; at night, it may be mainly sucrose formed from
the mobilization of starch stored in the leaf. Plants are programmed
to maintain a steady rate of assimilate translocation over an entire
24-hour period but how this is regulated is unclear.

Partitioning – sink strength
Partitioning of assimilate among sinks is also poorly understood.
One of the more significant factors determining the direction of
translocation is sink strength, which is determined by its size and
how active it is. A seed is a good example.

A high flow of food to a seed is required as it goes through its
phase of rapid increase in size. As its expansion slows, less food is
needed and the flow rate towards it moderates. A major influence
on the speed of movement of food towards the growing seed is how
quickly the food is unloaded from phloem into the seed. If the seed
is growing quickly, food is directed to it even at the expense of other,
more slow-growing, organs along the same route of translocation.

Sink strength is a powerful force in determining the direction
and speed of flow in phloem. But how is the need for more or less
assimilate transmitted to the source?

Phloem loading and unloading
As already noted, phloem translocation is pressure-driven. The loading of sucrose into phloem in the leaf generates a surge of water into

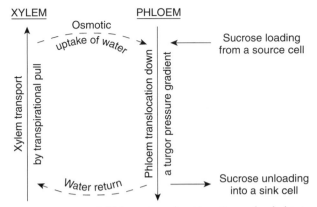

FIGURE 6 Phloem translocation. Sugar, loaded into sieve elements from a **source**, causes an osmotic surge of water from xylem into phloem and a build-up of turgor pressure in sieve tubes. Elsewhere, sugar is removed from the phloem at a **sink**, lowering turgor pressure there. The pressure gradient created causes water to flow from source to sink, carrying with it dissolved sugar. At the sink, excess water moves out of the phloem into the xylem.

the leaf phloem, creating a high turgor pressure which forces fluid along sieve tubes, out of the leaf. At a sink the opposite occurs. Food is unloaded into the sink, which lowers phloem pressure. Between source and sink, a pressure gradient is established which is more or less steep (i.e. stronger or weaker) depending on the strength of the sink. What is not understood is how changes to the steepness of the gradient along phloem sieve tubes are sensed at the source, but it appears to be a very sensitive feedback system which plays a major role in partitioning food materials among a wide array of competing sinks (Figure 6).

SUMMARY

- Plants do have a circulatory system for moving water, food (photosynthate), and other substances between their organs and tissues
- The system is complex and is controlled in such a way as to deliver balanced mixtures of nutrients to all parts of the plant body
- It is efficient, fast, and responsive to the changing needs of all plants, from the smallest to the largest, as they go through their life cycles

- In the case of the distribution of food throughout the plant, it is a pressure-driven system (see Box 6) through sieve tubes located in phloem tissue, originating at **sources** and ending at **sinks**
- We are beginning to have some idea of how food production (e.g. sucrose versus starch) in a leaf is regulated
- What is less clear is how distribution of food (allocation and partitioning among and between sinks) is controlled once it leaves a source.

The yields of many crops have been improved in recent years and a claim can be made, therefore, that it is possible to breed for a more favorable partitioning of available food towards areas of the plant which will be harvested. Some success has also been achieved recently in breeding for increased photosynthesis in leaves so that the total amount of food available for translocation out of leaves is greater. These successes suggest that a greater understanding of how allocation and partitioning are controlled is both possible and important.

BIBLIOGRAPHY: PART II

Barthlott, W., Porembski, S., Fischer, E. and Gemmel, B. (1998). First protozoa-trapping plant found. *Nature*, **392**, 447. *A plant which, like those that are insectivorous, captures protozoa and digests them for their nitrogen.*

Boonman, A., Anten, N. P. R., Dueck, T. A. *et al.* (2006). Functional significance of shade-induced leaf senescence in dense canopies: an experimental test using transgenic tobacco. *The American Naturalist*, **168**(5), 597–607. *The authors demonstrate that plants growing in dense stands benefit from shedding lower, shaded leaves, which can no longer photosynthesize efficiently, yet take energy to maintain.*

Crofts, A. Peter Mitchell (1920–1992). *A commentary on Mitchell's life and accomplishments, especially in regard to the chemiosmotic hypothesis. See:* www.life.illinois.edu/crofts (accessed July 18, 2010).

Dittmer, H. J. (1937). A quantitative study of the roots and root hairs of a winter rye plant (*Secale cereale*). *American Journal of Botany*, **24**, 417–20. *Grown under ideal conditions, this rye plant produced an astonishing amount of root.*

Epstein, E. (1972). *Mineral Nutrition of Plants: Principles and Perspectives*. New York: Wiley and Sons. *Comprehensive coverage of all aspects of plant mineral nutrition, including nitrogen fixation.*

Flowers, T. J. and Yeo, A. R. (1992). *Solute Transport in Plants*. London: Blackie. *Advanced coverage, to the date of publication, of all aspects of solute transport from the external substrate throughout the plant and its cells.*

Govindarajulu, M., Pfeffer, P. E., Jin, H. *et al.* (2005). Nitrogen transfer in the arbuscular mycorrhizal symbiosis. *Nature*, **435**, 819–23. *How some plant–fungi symbioses operate to transfer nitrogen from the soil to the plant via the fungus.*

Hales, S. (1961). *Vegetable Staticks [1727].* London: Oldbourne Press. *A re-issue of Hales's 1727 book which deals primarily with his experiments on the movement of water through the plant from the roots to the leaves and beyond.*

Hewitt, E. J. and Smith, T. A. (1975). *Plant Mineral Nutrition.* London: The English Universities Press Ltd. *Has an account of the history of plant nutrition after Van Helmont, including the important contribution of von Sachs' hydroponic method.*

Holderness, B. A. and Turner, M. (eds) (1991). *Land, Labour and Agriculture, 1700–1920.* London: The Hambledon Press. *The authors provide information surrounding the controversy between Liebig and Lawes and Gilbert regarding the nitrogen requirements of crops.*

Lewington, A. (1990). *Plants for People.* New York: Oxford University Press. *Information especially about the many ways soyabean can be processed to make different foods.*

Lewis, O. A. M. (1986). *Plants and Nitrogen. Studies in Biology*, No. 166. London: Arnold. *Information about many aspects of nitrogen assimilation and the nitrogen cycle.*

McCaskill, A. and Turgeon, R. (2007). Phloem loading in *Verbascum phoeniceum* L. depends on the synthesis of raffinose-family oligosaccharides. *Proceedings of the National Academy of Sciences USA*, **104**, 19619–24. *Overview of the early steps in phloem translocation from a leaf.*

Nelson, N. (1994). Energizing porters by proton-motive force. *Journal of Experimental Biology*, **196**, 7–13. *Review of the systems which carry out transport across cell membranes.*

Rachmilevitch, S., Cousins, A. B. and Bloom, A. J. (2004). Nitrate assimilation in plant shoots depends on photorespiration. *Proceedings of the National Academy of Sciences USA*, **101**, 11506–10. *Evidence of the essential role of photorespiration in nitrate assimilation by plants and why breeding crops with lower photorespiration may be counterproductive, therefore.*

Rothamsted Research. (2006). Guide to the classical and other long-term experiments, data sets and sample archives. www.rothamsted.ac.uk (accessed July 18, 2010).

Tudge, C. (1988). *The Environment of Life.* Oxford: Oxford University Press. *The importance of pulses to the human diet, and nitrogen fixation.*

Virgil's (2009). *Georgics: A Poem of the Land.* K. Johnson (translator). London: Penguin Classics.

Part III   Growth and development

A plant goes through a highly organized series of stages as it progresses through its life cycle. A fertilized egg, the zygote, divides repeatedly, differentiating into a wide array of tissues and organs, giving rise to a mature plant.

In Part III, chapters on how plants respond to cues from their environment (changes in **light**, **temperature**, and **gravity**) as they **grow** and **develop** are set beside discussions of how they sense **day** and **night**, respond to the changing **seasons**, attract pollinators with **color**, **fragrances**, and **flavors**, and use **dormancy** as a survival strategy.

7 Growth: the long and the short of it

INTRODUCTION

The fastest growing trees are the eucalypts, one type of which, found in New Guinea, has been known to add nearly 8 m to its height in 1 year. Even these "sprinters" are eclipsed by giant bamboo, which can grow over 1 metre a day, 30 m in under 3 months.

At the other extreme, a Sitka spruce found at the tree limit in the Arctic had one of the slowest growth rates on record. From measurements of the annual growth rings in the trunk it was estimated to be about 100 years old yet was only 28 cm tall.

The total growth of which some plants are capable in a lifetime is startling. One of the largest giant redwood trees found had a wood volume of more than 1500 m^3 and weighed over 1000 tonnes. Since the seed of the giant redwood weighs less than 0.005 g, the weight increase over the lifetime of this specimen was more than 250 billion times. Large trees like these can live for more than 4000 years, illustrating that plants often combine in their bodies tissues of great antiquity with others that are still youthful, producing new leaves, shoots, roots, fruits, and seeds.

CONTROL OF DEVELOPMENT AND GROWTH FORM

In animals, organs develop very early in life and become an integral part of the whole organism without which it cannot function. In plants, many organs are programmed to be created and discarded, repeatedly, over a lifetime, some of which often come and go, seasonally.

GENETIC CONTROL

All living things are programmed to take on particular shapes and growth forms. In plants, this genetic program is remarkably flexible, one example of which is the ability of plants to produce roots under special circumstances, a power made use of by gardeners when multiplying plants as cuttings. Roots are not normally formed directly on stems and even less often on leaves, yet a severed willow branch, when stuck in the ground, will often "strike," producing roots directly from the cut end. Fence posts made from the tropical gombo-limbo tree may quickly produce roots and leaves and in a few years grow into lines of trees along the edges of fields. Begonia leaves that come in contact with the soil for long enough may produce roots directly from the ribs of the leaf.

Activities such as these must be under some control which is genetic, at least in part. That petunias look alike but different from oak trees is due, in some measure, to differences in the genetic make-up of the two species.

CONTROL BEYOND GENETIC

In addition to genetic control of shape and form there is another kind of control responsive to the immediate environment of an organism.

Plants are not so rigidly genetically programmed that they flower at precisely the same time each year. Variations in temperature and moisture, for example, may cause plants to slow down or speed up their seasonal programs of growth and development. Flexibility is built in to allow the plant to respond to variations in its environment. The contrast between giant bamboo in a warm climate with Sitka spruce in the cold Arctic is one dramatic illustration of the variability possible.

This kind of flexibility is found for other growth responses as well. Somewhere, orders have not only to be issued for such things as more roots, shoots, or leaves to appear and fruits and leaves to fall in season but also for controlling the rates at which these activities occur.

PLANT GROWTH SUBSTANCES

Hints that plants have growth controls beyond genetic come from some familiar observations. We have all seen house plants bend towards a directional source of light. If a growing plant is laid on its side, in a few hours its branches will begin bending upwards, roots downwards, in response to gravity.

Theophil Ciesielski

The first recorded scientific observations that plants control their growth came in the 1870s.

- **Theophil Ciesielski**, a Polish scientist, found that when he removed the terminal cap from a root, the rest of the root no longer responded to gravity
- Ciesielski concluded that a stimulus was produced in the cap which caused a root, when laid horizontally, to curve downwards as it grew. He suggested that the stimulus caused the upper side of the root to grow faster than the lower.

We now know that Ciesielski was correct: the upper side does grow faster in a root laid horizontally than does the lower side, causing downward curvature. This is due both to a speeding up of growth on the upper side of the root and a slowing down on the lower side. What this stimulus was and how it had its effect was not understood in Ciesielski's day.

Charles and Francis Darwin – phototropism and gravitropism

The first observations that some substance or substances in plants affected growth were made by **Charles Darwin** and his son, **Francis Darwin**, during a study of the curvature of plants towards light. Working with seedlings of canary grass, the Darwins found that:

- if the very tip of a seedling shoot was covered with a tiny cap of blackened glass (to exclude light) the plant no longer bent towards light directed at it from one side;
- if, on the other hand, the seedling was buried in black sand so that *only* the tip was exposed, the *whole plant* bent toward the light;
- the Darwins also found that if even just the last 3 mm of the tip of a seedling were removed, no curvature toward light occurred.

The Darwins reported their conclusions in a book, *The Power of Movement in Plants*, in 1881, concluding that:

> when seedlings are freely exposed to a lateral light, some influence is transmitted from the upper to the lower part, causing the latter to bend.

In other words, the very tip of a canary grass seedling seemed capable of sensing the direction of light and passing on information to the rest of the seedling below, causing it to grow in the direction of light, a result reminiscent of that by Ciesielski with root caps and gravity. Only many years later was the nature of these influences on growth fully worked out and given the names, **phototropism** and **gravitropism**, respectively.

The nature of the stimulus

By the 1920s, others had concluded that the stimulus produced at seedling tips was a chemical, but plants contain thousands of chemicals. Which one was responsible for transmitting the "curvature stimulus" from the tip to the rest of the seedling? Was it only one substance or several acting together?

Fritz Went

Fritz Went, a scientist working in the United States in the late 1920s, reasoned that it should be possible to restore the ability to curve towards light to a decapitated seedling by adding to it the chemical lost when the tip was removed.

- Went painstakingly extracted and separated chemicals from shoot tips and added them back to decapitated seedlings one at a time
- After much patient work Went found that there was, indeed, a substance that restored the ability of seedlings to curve
- The identity of the substance remained elusive until it was isolated in larger amounts, not from plants, however, but from an unexpected source – human urine!

Fritz Kögl and auxin

Fritz Kögl, working at the University of Utrecht in the 1930s, set about trying to identify the elusive "curvature compound" in plants.

- He found that human urine, of all things, had in it a chemical which, when added to decapitated seedlings, restored their ability to curve towards light
- From about 180 litres of urine, provided by a local hospital, Kögl isolated 40 mg of crystals of a chemical which, when dissolved in water and added to decapitated seedlings, powerfully restored an ability to curve towards light
- The substance in urine proved to be identical to that found earlier by Went in seedling tips
- Kögl called the compound **auxin** (the Greek word *auxein* = *to increase*)
- Soon, others showed that auxin was also responsible for the bending of roots towards and stems away from gravity when plants were laid on their side.

Auxin, it was quickly realized, was a substance which fit the definition of a **hormone**, a word used first by animal scientists.

Definition of a plant hormone

A widely used definition of a plant hormone is:

> An organic compound synthesized in one part of a plant and translocated to another part, where, in very low concentration, it causes a physiological response.

Long before Went's and Kögl's studies, it was known that animals produced compounds in small quantities in one place and transported them elsewhere in the body where they had their effect. For example, hormones produced in our brains move in the bloodstream to the sex organs where they influence reproduction.

Auxin, too, is produced in one location in a plant (e.g. a seedling tip) in tiny amounts and then moved to other locations (e.g. lower down the stem or even all the way to the roots) where it affects growth. Eventually, it was realized that auxin is produced by *all* plants.

THE MANY EFFECTS OF AUXIN

Shoot elongation

As long as the lead shoot of a plant remains intact its side branches grow more slowly. A clear example of this is the so-called "Christmas tree effect."

Branches of conifers used as Christmas trees are arranged in a distinct way; short near the apex of the tree, becoming ever longer from the top, down. Many plants have this pyramid-shaped growth form as long as the lead bud remains intact. Side branches begin to grow more rapidly if, for some reason, the apex is damaged or lost, through the attentions of browsing animals or from disease, for instance. Gardeners and horticulturalists remove lead buds from their shrubs so that the plants will grow more "bushy" rather than tall and slender as is more likely to happen if the lead bud remains on a plant.

It is the auxin produced in the lead bud and transported down the main stem which inhibits the outgrowth of branches. As soon as this source of auxin is removed, branches grow faster.

Root initiation

At about the same time as the **branch-inhibiting** function of auxin was discovered, it was also shown to have **root-forming** activity. The commercial value of this was quickly recognized and put to use.

Cuttings of a wide variety of plants will grow roots directly from newly cut surfaces if auxin is added. Leaves, pieces of stem or root, or even bulb scales, if treated with auxin, can be induced to produce roots in places where roots grow only slowly, if at all, in a normal plant. Dipping the cut surfaces of cuttings into auxin solutions or powders has become standard horticultural practice. There are now dozens of such preparations available to gardeners.

Fruiting

Another property of auxins of economic importance is their ability to promote fruit formation without pollination (**parthenocarpy**) when added to certain plants. Tomatoes, for instance, are usually grown commercially in greenhouses where there are few insects and no wind to aid in cross pollination. Tomato growers, therefore, resort to spraying auxin on their plants to induce fruiting and avoid the slow task of pollinating each flower by hand.

Although apple and pear trees do not need auxin application to fruit abundantly, growers of these crops have used auxin for other reasons. A major source of loss in apple and pear crops is premature **fruit drop**; from a quarter to a half of a crop may be lost this way. Growers used to be faced with either harvesting before the best quality was attained or else risking a heavy premature fall. Spraying auxin onto apple or pear trees can delay the fall of fruit, increasing the harvest.

Other uses of natural and artificial auxins
Other commercial applications of auxins take advantage of their activity in inhibiting, rather than promoting, growth in certain cases. For example, potato tubers can be treated with auxin, preventing them from sprouting in storage; in this way tubers can be kept longer. Artificial auxins, such as 2,4-D (2,4-dichlorophenoxyacetic acid), are used to control broad-leafed weeds while having no effect on grasses. Under the right conditions, 2,4-D is used on sugarcane and maize crops as well as golf courses and lawns to control common weeds.

PLANT GROWTH SUBSTANCES ARE NOT LIKE ANIMAL HORMONES
This short list of the effects of auxins illustrates that, while the activities of substances that influence plants parallel those of animal hormones up to a point, the parallel is not complete or exact. Hence, the need to give them a different name and look at them differently from their animal counterparts.

Whereas an animal hormone is likely to control a single process or function, a **plant growth substance** influences a plant in many ways. It may stimulate growth or, under other conditions, it may inhibit growth. Animal hormones often have precise targets and have no effect anywhere else in the body. The plant equivalents have general effects on many aspects of growth and development all over the plant body.

OTHER PLANT GROWTH SUBSTANCES

For two decades following its discovery in the late 1920s, auxin (or **indole acetic acid**, IAA, as it was eventually shown to be) was the only natural plant growth substance thought to exist. Repeated attempts were made to show that auxin could control all plant growth and development, but it became increasingly obvious that there could not possibly be only one growth substance in plants.

Tall and dwarf peas

One reason for this conviction had to do with tall and dwarf pea varieties. Auxin was known to increase growth so it seemed logical to assume that adding it to dwarf peas would make them grow tall – not so. Auxin had no such effect. Ironically, the answer to this dilemma was already at hand when the question was raised.

GIBBERELLINS

Eiichi Kurosawa

At about the same time as auxin was being identified, other discoveries were being made in Japan.

A young scientist, **Eiichi Kurosawa**, had been investigating a serious problem in *the* major crop in Japan.

- Rice plants infected with the fungus *Gibberella* grow tall and more spindly than normal
- This excessive growth weakens the stems of the rice plants, which then "lodge" (collapse) more easily in heavy rain or high wind
- This, in turn, reduces the eventual yield as it does in any crop
- The Japanese have given this condition in rice the name "foolish seedling disease"

- Kurosawa found he could make the stems of rice plants grow long and spindly without infecting them with the *Gibberella* fungus if he added to the leaves of rice plants the nutritious broth in which he had grown the fungus. Clearly, the fungus produced a substance which leached into the broth
- By the late 1930s, Japanese chemists had isolated an active, crystalline substance which they called **gibberellin A**
- When it was found that plants also produced compounds very similar to gibberellin A it was realized that here was a second possible growth substance to set alongside auxin.

Effects of gibberellins

Well over 100 gibberellins have now been discovered, all of which have much the same chemical structure. One of their most pronounced effects is to change the **rate** of growth, including dwarf stems. Dwarf peas *can* grow tall; all they lack is enough of the right kind of gibberellin to stimulate their shoots to elongate more rapidly. Addition of gibberellin to a cabbage plant converts the "head," which is really a dwarf stalk, into a stem 1.8–2.4 m tall. Plants like sugar beet, which form a "rosette" of leaves close to the ground, can be made to "bolt" to great heights by a gibberellin treatment.

Incidentally, treating naturally tall plants with gibberellin does not affect their growth. They already contain enough of the growth substance and do not benefit from having more added.

Other gibberellin effects

The dramatic effects of gibberellins do not begin and end with dwarf plant growth. Another event in the life of many plants is **dormancy**, a topic important enough to have a chapter of its own (Chapter 10).

In temperate climates in late summer, deciduous plants produce overwintering buds which grow out the following spring. These dormant buds form even though the air temperature at the time they are produced by the plant is still high enough for the rest of the plant to continue growing for many weeks. Seeds, too, often show dormancy.

Many weeds will germinate only if the ground in which they are lying is disturbed (by cultivation or by burrowing animals, for example) and they find themselves exposed to light after being buried, perhaps, for years. Treatment of dormant seeds such as these with gibberellin in the dark causes them to germinate. Treatment of dormant buds with gibberellin likewise stimulates them to grow immediately.

Gibberellins are widespread

Gibberellins are found in many organisms. Each species of plant has at least a few but they are also common among fungi and bacteria. All parts of plants contain gibberellin, with the highest amounts found in seeds. Young tissues have more than old ones but, in general, they are concentrated in the most active parts of a plant.

The discovery of this second type of growth substance in plants was just the beginning of the search for growth-active compounds to add to auxin.

CYTOKININ: THE "WOUND HORMONE"

As early as 1913 it was shown that when plants were wounded, new repair tissue quickly formed at the damaged surface, rather as a "scab" forms in animals. Repair was prevented by washing wounds with running water immediately after damage occurred. The obvious conclusion was that water washed away something essential for healing. The unknown substance was called a **wound hormone** but its identity remained elusive.

Of course, the new tissue for wound healing is formed from new cells which replace those damaged during cutting. Thus, the idea arose that a place to look for a wound hormone would be where new cells were routinely produced. One obvious place to search was the ovule during seed formation. Here, not only is the new embryo formed but also all the other parts commonly found in a mature seed.

In the 1960s, a natural compound with a powerful influence on cell division was isolated and purified from developing maize seeds.

This class of plant growth substances became known as **cytokinins** (*cyto* = cell; *kinein* = movement or growth), a name which describes the ability of these substances to stimulate plants to rapid production of new cells. They were quickly found to occur in all plants, in fungi, and in bacteria.

ETHYLENE: THE FIRST GASEOUS HORMONE

The most unlikely plant growth substance is a compound known for centuries to affect plants but not designated a plant growth substance until the 1960s.

The Chinese knew long ago that fruits would ripen more quickly in a room where incense was burning. Puerto Rican pineapple and Philippine mango growers built bonfires to produce smoke to synchronize flowering in their crops. Illuminating gas (for example, coal gas, once used to light streets, homes, and offices, especially in Europe) leaking out of pipes caused leaves to fall off shade trees in certain German cities (reported as early as 1864).

By the 1930s, there was a general but still vague understanding that all these apparently unconnected effects were caused by one substance, the gas **ethylene**. That something which was a gas at normal temperatures could be a plant growth substance took much longer to gain acceptance. We now know that most, if not all, plants and plant parts produce ethylene. Mostly, the amounts released are small but even then can affect plant growth and development.

Ethylene was the first gaseous compound found to act as a hormone.

The 'climacteric' in fruits

The most dramatic effect of ethylene is on the speed with which certain fruits ripen. Apples, pears, tomatoes, and bananas, for example, produce much increased amounts of ethylene just as their fruits begin to ripen. The extra ethylene hastens ripening, called the **climacteric**, and can lead to one fruit influencing the ripening of another. The old saying, "one rotten apple spoils the whole barrel," refers to the fact that if apples (or other "climacteric fruits") in a closed container have among them just one that is overripe, the excessive amount of ethylene produced by this one fruit, because ethylene is a gas, can spread throughout the container and accelerate the ripening of the rest. Spoilage of this kind can be minimized by refrigeration (but not in the case of bananas, which are temperature-sensitive for other reasons), since low temperature slows ethylene production, or by quickly removing the ethylene by increased ventilation.

Not all fruits produce excess ethylene. In grapes, cherries, and the citrus fruits, ethylene plays no part in ripening. Yet, as far as can be determined, all parts of all the flowering plants produce at least some ethylene, if not necessarily very much.

OTHER ROLES OF ETHYLENE – STRESS

Numerous **mechanical and other stress effects** on plants also increase ethylene production, including increased mechanical pressure on a leaf or stem; attack by fungi, bacteria, viruses, or insects; waterlogging of soil around roots; and drought soil conditions.

In seedling growth

As a young seedling pushes its way up to the surface of the soil, having just germinated from seed, it is still very weak. If it is blocked by hard, baked soil, for instance, it may not be able to force its way to the soil surface. It reacts to the barrier by producing more ethylene, which causes the stem to slow its upward growth but expand its thickness. The sturdier stem provides the bulk needed for a more determined upward push.

In plant defense

Overproduction of ethylene by a plant at the point of attack by insects, bacteria, fungi, or viruses can lead to the death of the plant tissue immediately around the invasion site (**necrosis**). Brown spots on leaves under attack by insects or diseases may be a sign of such planned death of cells and tissues. Ethylene production helps speed up this response, which provides a measure of defense against the further spread of an attack.

Disease organisms are less likely to spread if the tissue surrounding them is already dead. They become isolated in a ring of dead tissue from which they cannot escape to invade the surrounding, living, tissues. Of course, this response does not always succeed in stopping the advance of an attack but may slow it enough to give other defense strategies time to work.

The ethylene may also act by directly slowing down the growth of the invading disease organism itself (see Chapter 13 for a discussion of plant defense strategies).

In speeding up senescence
Thus, ethylene is a growth substance which helps the plant to react to various external stresses, the end result being a speeding up of maturation in affected tissues. This view of the role of ethylene is further confirmed by the fact that its production is often increased in flowers as they fade. After flowers have served their function they wither away quickly in most plants, a process hastened by increased ethylene production in petals (see Chapter 14).

In speeding up flower formation and seed germination
Ethylene also speeds up flower formation in some plants and promotes seed germination in others. It is not just an inhibitor but has multiple effects on growth. In this, it resembles all other plant growth substances (Figure 7).

ABSCISIC ACID – THE BRAKE PEDAL
Attack by predators and diseases are not the only stresses to which plants are subjected. In the normal course of a day in the hot sun, plants may find themselves short of water or be the victims of high temperatures from which they cannot escape by moving into the shade. Conversely, a plant might find itself exposed to sudden cold

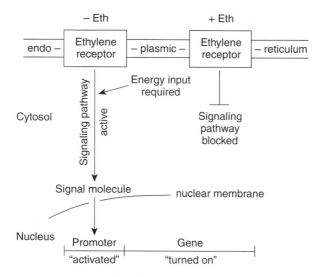

FIGURE 7 One example of how a plant hormone acts. There are receptors in the endoplasmic reticulum of cells which recognize ethylene (Eth). In the absence of the hormone (– Eth), a signaling pathway in the cell is active, a signal molecule is transported into the nucleus, turning on specific gene(s). If ethylene appears in the cell (+ Eth), it binds to its membrane receptor, which leads to the blocking of the signaling pathway: the gene(s) linked to the pathway are turned off (after Hopkins and Hüner, 2009).

which must be endured for longer (over a winter) or shorter (overnight) periods of time.

Search for an inhibitor

A group of investigators in the UK, searching for an explanation of why sycamore trees form winter buds long before the end of summer, asked how these plants "know" winter is approaching when the weather is still warm? Others in the United States and in New Zealand were trying to find out what causes fruits to fall when they are ripe. The three groups independently came to the same conclusion at about the same time: that plants have a growth substance, **abscisic acid** (ABA), named because of its influence on fruit and leaf fall, or **abscision**, which influences many aspects of plant growth and development.

Roles of abscisic acid

If a plant lacks water, it increases production of ABA, which causes the stomata in leaves to close (see Chapter 3). In this way a plant conserves whatever water it still has. In late summer more ABA is produced, which aids in the production of winter buds in many trees and perennials. ABA produced in seeds can cause them to become dormant (see Chapter 10) and remain so until the level of ABA in the seeds declines some time after they have been shed from the parent plant. Higher or lower than normal temperatures can lead to the production of increased amounts of ABA in all parts of the plant body which causes, in turn, the growth of all parts of the plant to slow down. When normal temperatures return, the level of ABA in the plant decreases and growth again speeds up.

ABA, then, acts something like the brake on a vehicle. When conditions surrounding the plant are not optimum, or in preparation to survive adverse conditions in the future (as in the case of seed dormancy or winter bud formation), ABA acts to slow the pace of, or stop entirely, normal life processes until more favorable conditions return.

BOX 7. BRASSINOSTEROIDS AND JASMONATES

Steroid hormones have long been known to play key roles in animals (e.g. the sex hormones: estrogen, androgen, progestin). They were first recognized in plants in the early 1970s when compounds in organic solvent extracts from Brassica napus *pollen (hence their name,* **brassins***) were shown to stimulate cell division, cell elongation, increased total biomass, and seed yield. The first bioactive compound purified from pollen extracts was* **brassinolide***, which remains the most active member of this group of plant growth substances. By the 1990s, a family of* **brassinosteroids** *(BR) had been isolated from many plant species. They have been shown to interact with other plant growth substances, especially auxin, in regulating many aspects of plant*

BOX 7. (cont.)

development, including shoot, root, and leaf growth, xylem and phloem differentiation, fertility, seed germination, senescence, and cell death.

Many pathways in plants lead to BR synthesis, all of which are linked to those involved in the production of gibberellins and ABA. They act by attaching to cell membranes, which triggers a long sequence of biochemical steps ending in the nucleus of a cell. There, the signals initiated outside the nucleus cause the activation of BR-sensitive genes involved in plant development.

The **jasmonates**, principally **jasmonic acid** (JA)and **methyl jasmonate**, are a group of compounds which were first recognized because of their involvement in insect and disease resistance in plants. Now, they are known to modulate a wide variety of other physiological processes, including seed and pollen germination, protein storage in tubers, root development, fruit ripening (with ethylene), and proteins involved in photosynthesis.

JA is synthesized from the fatty acid, linolenic acid. The latter is abundant in cell membranes, which are the source of it for JA biosynthesis. JA levels in plants vary with tissue and cell type, developmental stage, and environmental stimulus. For example, JA levels rise steeply in response to insect and pathogen damage, triggering activation of a host of genes, then, the synthesis of proteins involved in plant defense. High JA levels are also found in flowers, fruits, and chloroplasts in the light. JA also increases rapidly in tendrils as they coil around objects, and when plants are wounded.

Among the many components of plant defense induced by JA are inhibitors which interfere with digestion in insects. Insect feeding causes a plant to increase its content of these inhibitors which, when they enter the herbivore's digestive system, prevent the breakdown of proteins ingested from the plant. This leads to a slowing of insect growth and development.

SUMMARY

- A plant's genetic program may be the ultimate control of growth and development, but growth substances are a necessary additional level of day-to-day response to the vagaries of climate and weather. Without them plants would be "blind" to variations in their environment
- The long and the short of it is that plants contain a number of growth substances which, either separately or together, guide the progress of all aspects of their growth and development in response to variations in their environment
- The five types of these compounds described above (auxins, gibberellins, cytokinins, ethylene, and abscisic acid) are not the only ones
- More recent investigations of substances in plants called brassinosteroids and jasmonates are cases in point (see Box 7). Others will surely be discovered as more detail is learned about plant growth.

8 The time of their lives

INTRODUCTION

Plants growing in those areas of the world where there are definite seasons synchronize their activities to suit regular variations in the weather. It makes little sense for a seed in temperate regions to begin germinating as soon as it is produced if it is late summer. Onset of winter would likely freeze and kill the fragile seedling before it establishes itself. A tree or shrub faced with an approaching cold season begins, in advance, to make provision to protect tender growing points and leaves within buds (see Chapter 9).

In another example, annual herbs develop vegetatively before diverting their energies to forming flowers, fruits, and seeds. Leaves are most often produced first to supply enough food reserves through photosynthesis before energy-consuming tasks like reproduction begin. Such sequences of events are timed to occur in synchrony with the seasons.

For these, and many other reasons, plants have evolved ways to sense the passage of time; to measure the lengths of days and the onset of seasons. Rhythms in plant life are linked to such cycles in their environment through a number of remarkably exact time-measuring mechanisms.

DAILY AND SEASONAL RHYTHMS IN PLANT LIFE

Sleep movements

It has been recognized for a long time that rhythms in plants exist. **Androsthenes**, historian to Alexander the Great, was one of the first to observe, over 2300 years ago, that the leaves of some plants adopt different positions during the day than at night. In some, leaves come together at night, unfolding again during the day. Others have leaves which point downwards during the night, at an upward angle during the day. The South American Rain Tree folds up its leaves not only at night but also in cloudy weather, just as many sun-sensitive flowers do.

These so-called "sleep" (**nyctinastic**) movements are common not just among leaves but also in flowers. **Carolus Linnaeus**, who established the system of classifying plants by giving them binomial Latin names, developed a "floral clock" based on opening and closing of various flowers, which was reputedly accurate to within half an hour.

Synchronized plant growth and development

One of the most accurate plant time-keepers so far discovered is the Malayan evergreen bush, Shrubby Simpoh, which flowers every day of its life of half a century or more once it has reached full maturity. Its flower buds open at 3 am; its petals fall at 4 pm the same day; fruits ripen in precisely 5 weeks; they split open to release seed at 3 am of the 36th day after flower buds first open.

These examples illustrate that, in plants, there are daily and seasonal rhythms. In addition to flower formation and the sleep movements of leaves and petals there are the regular variations in such things as growth rates of various organs (leaves, stems, roots, flowers), which may grow faster or slower at different times of day; fluctuations in levels of growth substances which increase and decrease on a 24-hour cycle; the daily opening and closing of stomata in leaves; the release of scents by plants correlated with nectar production; changes in the rate at which photosynthesis takes place; and variations in respiration, to name a few of the many known.

THE BIOLOGICAL CLOCK

It was sleep movements of leaves that, historically, led to the earliest clear observation of rhythms in plants and the fact that plants have something like a "clock" which they use to keep time.

Jean Jacques d'Ortour de Mairan

More than two and a half centuries ago the French astronomer, **Jean Jacques d'Ortour de Mairan**, asked a crucial question about the sleep movements of plant leaves. He wondered whether they were caused by changes in the environment around a plant (regular day–night changes, for example) or whether a plant had an internal time-measuring mechanism.

- Using the *Mimosa* plant, which he knew already to have very sensitive and predictable daily leaf movements, de Mairan noted changes in position of leaves in bright light compared to deep shade
- He found that the leaves of *Mimosa* continued to open and close, daily, even in continuous, near complete darkness, from which he concluded that there was a "clock" *inside* the plant guiding sleep movements of leaves, independently of the environment.

We know now that de Mairan was correct; plants do have internal, independent clocks. What he missed is that regular, predictable changes in the environment surrounding a plant *do* influence the way in which the internal clock operates. Over 200 more years would pass, however, before the idea that plants had an internal system for telling time would take hold. Even today, the links between biological clocks and the environment are by no means fully understood.

We now suspect that all living things have internal clocks. For example, humans have rhythms in sleep, in the ability to stay alert and make complex decisions, in varying hormone levels, heart rates, body temperature, excretion of urine, sensitivity to drugs, births, and deaths (both of which occur most frequently between 2 am and 7 am and least often between 2 pm and 7 pm), and a host of other regular bodily functions.

Erwin Buenning, Kurt Stern, Rose Stoppel, and factor X

The most important breakthrough in early understanding of clocks in plants came in the 1920s.

- In Frankfurt, two scientist, **Erwin Buenning** and **Kurt Stern**, were investigating leaf movements in the common bean plant which, like

Mimosa used by de Mairan two centuries earlier, expose their leaf surfaces to the sun by day and fold them vertically to a sleep position at night

- They were aided by **Rose Stoppel**, who found that when she measured the movements in a darkened room at constant temperature and humidity leaves were always at their maximum sleep position between 3 am and 4 am each day.

Although this result pointed to the same conclusion as de Mairan, that plants have an internal clock, Stoppel refused to believe that the plants alone were capable of keeping such accurate time. She was convinced that some outside influence somehow reset the clock each day. She had no idea what this influence was and called it simply, **factor X**, reasoning that it had to be something other than light, temperature, or humidity, all of which she thought were under her control; in this she was mistaken.

Red and far-red light

Stoppel was methodical. One of her habits was to water her bean plants at the same time each day. In order to see to do this in the darkened room where the plants were housed she used a flashlight covered with red paper. The belief at that time was that red light had no effect on plants. For Stoppel, this was the kind of serendipitous error that all scientists wish they might make once in their lifetime, an error leading to an unexpected, important discovery.

Stoppel eventually returned to her home base in Hamburg without discovering the identity of factor X. Soon after, Buenning and Stern moved the bean experiment from where it had been located to Stern's potato cellar where there was better temperature control.

- Like Stoppel, they found that the maximum sleep position of the bean leaves always occurred at the same time each day
- What surprised them was that the maximum did not occur between 3 am and 4 am as Stoppel had found but about 8 hours later, between 10 am and noon.

The truth dawns

Buenning and Stern quickly recognized that the key to the dilemma was the dim red light they, and Stoppel, used in their daily routine of watering plants.

- The potato cellar used for growing the beans was far away from the laboratory where they worked during the day so they waited until late afternoon before carrying out this task; Stoppel had done it in the morning, 8 hours earlier than Buenning and Stern
- The daily brief exposure to dim red light was enough to reset the bean plants' clock so that 16 hours later the leaves folded to their maximum sleep position
- The brief exposure to dim red light was interpreted by the bean plants as a signal that dawn was breaking; the transition from dark to light each day reset the biological clock controlling sleep movements.

In regions away from the equator, dawn comes at a slightly different time each day, year round. Resetting the clock in response to a reference point such as dawn allows a plant to measure time. Of course, Stoppel, Buenning, and Stern did not vary their simulated "dawn." Therefore, the next point of maximum folding of the leaves in sleep was always at the same relative times in their experiments.

Thus, **red light** was Stoppel's **factor X**. Plants are indeed affected by red light, a discovery which led to one of the most imaginative and intuitive biological investigations ever undertaken, as we shall see in the next chapter.

CIRCADIAN RHYTHMS

Buenning and Stern went on to make another unexpected, important discovery.

Free-running rhythms and entrainment

- Leaf sleep rhythms in their bean plants grown in continuous darkness no longer kept to a strict 24-hour cycle
- The cycle became 25.5 hour and was soon completely out of phase with the cycle of light and dark outside their potato cellar; the rhythm had become **free-running**

- A 24-hour cycle could be reimposed, **entrained**, by once again briefly exposing the plants to dim red light at the same time each day. Whenever that was done, the maximum sleep position of leaves was reached, as before, 16 hours later
- Because the free-running rhythms were close to 24 hours in duration, but not exactly so, they became known as **circadian** (*circa* = about; *diem* = a day) **rhythms**
- In constant conditions, such as continuous darkness or light, *all* rhythms linked to regular day and night cycles, not just sleep movements of bean leaves, drift away from an exact 24-hour cycle
- They move to some value which is either shorter (21 or 22 hours) or longer (25 or 26 hours) than a regular day
- These natural circadian rhythms can be brought back onto a 24-hour cycle by exposure to some regular, repeating signal such as a dark to light transition at dawn or, in some cases, dusk.

Clock independent of temperature

Furthermore, the internal clock which measures passage of time is largely independent of temperature. A clock of any kind would be useless if the rate at which it operated changed as temperature varied. An essential characteristic of a clock is that it keep time no matter what is going on around it. This is no less true for the biological variety than for any other.

Zeitgebers – time-givers

Signals like dawn and dusk have been given the name **Zeitgeber** (the German word for *time-giver*). Unless reinforced by regular Zeitgebers, circadian rhythms will drift to some cycle length other than the environmental one.

The light–dark Zeitgeber

After a while if a light/dark stimulus is not given, even the free-running rhythm will become progressively weaker and eventually peter out. Some stimulation from the environment is crucial to maintaining these rhythms even in their free-running form.

Resetting the clock using something like the transition from dark to light at dawn or light to dark at dusk makes sense since many rhythms are linked to day-time or night-time activities. Because away from the equator the lengths of day and night vary with the seasons, some signal linked to that fact is essential. For instance, it is important that leaves be in their fully open position in the middle of the day for maximum photosynthetic efficiency. To keep track of the start of the day, however, some mechanism is needed to inform the plant when the transition from dark to light is occurring so that the reaction of the leaves is synchronized with daylight hours. The light–dark transition Zeitgeber is widely used by plants to set internal clocks not just for leaf movement but for many other activities as well.

The tidal Zeitgeber

Light and dark are not the only Zeitgebers. In ocean tidal zones, marine animals and plants synchronize their feeding and reproductive activities to the ebb and flow of the tide. If removed to places without tides, some marine organisms will continue this imposed pattern of behavior as though the tide was still there. Eventually, as with any rhythm, the activity will peter out unless stimulation is provided.

From time to time, claims have been made that there are rhythms in addition to tides associated with the lunar month, but evidence is not free from alternative explanation.

Environmental factors not useful as Zeitgebers

There are many repetitive changes in environments other than light and dark and the tides. Wind velocity, temperature, light, and humidity levels all change periodically. But, in reality, only three of these cycles are regular enough for organisms to evolve mechanisms to take advantage of them. Changing temperature, wind velocity, light levels during the day (for instance, owing to changing cloud cover), and humidity are all too unpredictable to form the basis for regular reactions to them. Daily changes in light and dark, annual fluctuations in day length, and the tides, are the only environmental variations sufficiently reliable for use in setting and resetting clocks.

OTHER RHYTHMS

Circumnutation

Although they are not visible to the naked eye, since they occur rather slowly, some movements of parts of plants have regular cycles which are much shorter than a day. For example, as plant shoots lengthen they do not move ahead perfectly straight and steady; they sway in gentle, repetitive circles. Charles Darwin called this motion **circumnutation**. In peas, one complete cycle of this motion takes about 77 minutes. As leaves expand to their full size, they sway and rotate in repeatable patterns which take just a few minutes.

Rhythms longer than a day

There are also rhythms longer than 24 hours; 4-day rhythms occur in the growth and reproduction of some fungi. A few cycles of one or several years are also known. The reaction of bean plants to red light has already been dealt with in some detail but is not the same at all times of year. Beans are more sensitive to red light during the normal growing season than at other times of year. The flowering of some bamboo species have cycles of 30 or 40 years. One particular bamboo species native to the mountains of Jamaica carries the process of growing to a remarkable extreme: exactly 32 years after the plant germinates from seed it flowers once, then dies, a pattern independent of environment. When transplanted to another part of the world the plant still blossoms on schedule at the age of 32 years.

Multiple cycles

Many organisms have more than one free-running cycle. For instance, such things as periods of activity and calcium excretion in humans both have natural cycles 33 hours long whereas body temperature, urine volume, and release of potassium into the urine, all have free-running cycles of about 25 hours.

Why so many cycles?

The reason for rhythms with periodicities other than 1 day may be lost in the dim and distant past. For example, it is believed that two and a half billion years ago, the lunar month might have been 40–45 days long, with each day being only 5 hours in length. Others have suggested a 425-day year for our planet 600 million years ago.

Perhaps rhythms differing widely from a 24-hour period are evolutionary remnants of times when day lengths and other regularly cycling conditions were not as they are now. On the other hand, maybe some living things have just never quite got it right or are on different cycles for reasons we do not yet understand.

NATURE OF THE CLOCK(S)

By what mechanism do clocks control what plants do? Is there more than one kind of clock in an organism? Can they be different in different organisms?

Some progress is being made in answering questions like these through the isolation of mutants in which clocks are modified. In some species, mutants have appeared which have certain rhythms that are different from the normal in ways that are passed on to the next generation (see Box 8).

BOX 8. NATURE OF THE CIRCADIAN CLOCK

*Plants are highly sensitive to light, a key influence at all major developmental stages in their lives. They have groups of **photoreceptors** for detecting UVB, UVA, blue, red, and far-red light. Two of these sets of receptors, **cryptochromes** and **phytochromes**, are used to regulate circadian clocks, among many other physiological roles. Cryptochromes are blue light receptors; phytochromes respond to red and far-red light (see also Chapter 9). The former are located in cell nuclei; the latter are formed in the cytosol and remain there until converted to their active form when they move to the nucleus.*

BOX 8. (cont.)

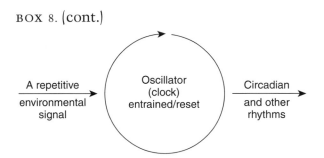

A repetitive environmental signal → Oscillator (clock) entrained/reset → Circadian and other rhythms

FIGURE 8 A biological clock mechanism comprises a reliable, repetitive signal from the environment (e.g. in plants, light versus dark or annual day length variations) which entrains/resets a central oscillator in each cell. The oscillator, in turn, controls the expression of many different biological rhythms.

Understanding how a circadian clock works in plants continues to be a major focus among plant researchers. A clock seems to consist of interconnected genes which activate and repress one another in a cyclical manner over a 24-hour period. Products of these genes rise and fall, successively, creating an **oscillator** *which conveys a sense of day and night length to all parts of the organism at all stages of development (Figure 8).*

Arabidopsis thaliana (thale cress) is a model plant used to isolate mutants with modified circadian clocks. In one set of studies, as an example, Arabidopsis *plants with normal versions of two genes (here called A and B) known to be involved in the clock were compared with plants with mutated forms of these genes. A faulty gene A caused the plant's circadian clock to speed up to a 20-hour cycle; mutations in gene B slowed down the clock to 27 hours. It was shown that the product of gene B degrades the product of gene A; this degradation is a vital step in maintaining the normal 24-hour functioning of the clock.*

Away from the equator, the constantly changing time of dawn and dusk over the year is sensed by the photoreceptors, cryptochromes, and phytochromes, an **input** *of information*

BOX 8. (cont.)

*used to set and reset the activities of the gene components of an oscillator, daily. Rhythmic **output** signals from the oscillator are linked to clock-controlled genes involved in diurnal or seasonal processes via complex signaling pathways. These additional layers of complexity need not all be entrained themselves to a strict 24-hour cycle; they could have different lengths (from minutes to days to months, etc.), yet take information about changes in day and night length and, hence, seasons, from the same oscillator.*

What genetic studies have not yet resolved are questions such as the following. Is there just one oscillator in each cell? If so, is it the same in all cells? Or, are there several oscillators in a cell linked to different physiological processes or developmental events? If more than one, are the types of oscillators the same in all cells or may they be different?

SEASONAL RHYTHMS

Finally, despite the concentration here on circadian rhythms and the establishment in plants of the existence of daily clock mechanisms, it is arguable that seasonal rhythms are by far the most important to plants overall.

At the beginning of this chapter, examples were given of the regular seasonal flowering of different types of common plants. Other long-term activities are such things as the production of buds, the development and fall of leaves, especially in deciduous trees, and the preparation for winter. Gardeners know well that activities such as production of buds or fall of leaves are not markedly affected by weather. A very cold snap may slow leaf or flower bud opening for a time but if bad weather persists buds will open eventually even if the flowers or leaves then freeze. Conversely, an unseasonably warm period can bring plants prematurely into leaf or flower.

Seasonal variations in the times of opening of leaf and flower buds are, however, usually small, often only a few days either way from year to year. Therefore, plants must have ways of setting rhythms which also accurately measure the onset of seasons. How this is achieved is the subject of the next two chapters.

SUMMARY

- Plants, which live their lives fully exposed to their environments, require a means to measure, in particular, changing day/night lengths and the onset of seasons
- Dawn and dusk transitions of light and dark are the most widely used Zeitgebers (time-givers) by plants, but the cycles of ocean tides across the globe are also sufficiently precise to be used by marine organisms to measure time
- Information from these environmental Zeitgebers is used to set and reset a circadian clock inherent in each living cell of an organism
- This clock is based on a genetic (gene) oscillator, entrained to a Zeitgeber (see Box 8)
- If Zeitgebers are removed (e.g. by keeping a plant in constant darkness or light), the clock is no longer circadian. It adopts a free-running rhythm either shorter or longer than 24 hours
- If a Zeitgeber is removed for long enough, even the free-running rhythm peters out; the rhythm reappears once the Zietgeber is restored
- Using the precise information provided through its circadian oscillator, a plant is able to maintain a wide variety of rhythms in its life, varying from a few minutes to, possibly, years in length.

9 A dash of seasoning

INTRODUCTION

Plants not only use day to night transition to time repeated functions from day to day (see Chapter 8) but also to measure the seasons of the year. Down the centuries humans have used leaf, flower, and fruit production by local plants as signals for seasonal activities such as when to begin planting crops or when to harvest. The agriculturalists Garner and Allard (more about whom shortly) expressed it this way:

> One of the characteristic features of plant growth outside the tropics is the marked tendency shown by various species to flower and fruit only at certain times of the year. This behaviour is so constant that certain plants come to be closely identified with each of the seasons, in the same way as the coming and going of migratory birds in spring and fall.

PATTERNS OF GROWTH

Away from the tropics
In temperate climates there is a pattern of plant growth arranged around a yearly period of relative inactivity, winter. The pattern is seen most clearly in annuals, which grow and reproduce during favorable weather and spend winter as seeds.

Other kinds of plants also have distinct patterns of growth. Herbaceous perennials produce annual stem growth which ceases well before winter; deciduous shrubs and trees lose their leaves in the fall but begin preparing for that event well in advance of winter. In all cases of periodic growth, some signal causes the plant to go from rapid growth to near complete shutdown in days or weeks.

In tropical regions

In the tropics, where variation in light and dark is absent or only small, plant responses are different. Some tropical trees are laws unto themselves. If they rest at all, losing leaves for a short time, they do so individually, not synchronized with their neighbors even when of the same species. One species may flower in a cycle of a little less than 1 year; another in a few months greater than an annual cycle; some do so continuously, often one branch at a time; yet others bloom only once every few years. This laissez-aller behavior is consistent with the congenial, nearly constant, environment typical of the tropics in which time-keeping is of lesser importance.

Species transplanted to the tropics

Plants that normally grow in stricter temperate climates can behave strangely when transplanted to the tropics. For example, pear trees planted in Indonesia became evergreen, with each bud having its own cycle of activity not synchronized with others on the same plant.

There is little evidence that plants are aware of the passing of a period of time as long as a year. In any case, we would find it difficult to show that a plant was measuring such a long span of time. Yet, undeniably, many plants do behave as though strongly linked to the seasons. Somehow, they are capable of marking these changes. What is it that they measure? The beginning of an answer to that question came through the study of flowering in plants.

MEASURING THE SEASONS

Julien Tournois: day and night length

In 1910, a doctoral student in Paris, **Julien Tournois**, began studying flowering in Japanese hops. A typical student:

- he was impatient to move his study ahead so he planted seed in February, 1911, instead of at the usual time later in the spring;
- to his surprise, his plants flowered in April, even though they were only 15 cm tall. Hops are usually leggy, twining plants which produce meters-long stems before blooming in July;

- when, the following year, plants started from seed in January flowered in March, Tournois began thinking that day length was affecting the timing of flowering. He wondered whether deliberately shortening the length of a day during a normal growing season would produce the same result;
- in April, 1912 and 1913, he sowed hops in his garden under glass. Those plants grown in natural cycles of day and night grew tall and spindly and produced nine pairs of leaves before flowering in early July, as expected;
- other individuals covered with dark screens for all but 6 hours a day grew just above ankle height, developed only three or four pairs of leaves and flowered around June 20th. Tournois concluded that: "in the Japanese hop, a decrease in day length during the normal growth period provokes some floral reproduction;"
- after seeing the same result with hemp, Tournois' conviction that his plants flowered when still young because they were exposed to short days from germination onwards became even stronger;
- what is more, he came to the additional, most important conclusion that it was the length of the *night*, not shortness of the *day*, that determined when his plants flowered.

With these deductions, Tournois was on his way to making a significant contribution to our understanding of the control of seasonal rhythms in plants, a promise which, sadly, would never be realized; he was killed in action in World War 1.

Wightman W. Garner and Harry A. Allard: Maryland Mammoth tobacco

This, and other early work on the control of flowering in plants, was unknown to two other investigators, **Wightman W. Garner** and **Harry A. Allard**, working at the United States Department of Agriculture near Washington, DC, shortly after World War 1. They questioned why the variety of tobacco they were researching, Maryland Mammoth, flowered so late in the season that its seed failed to mature before the onset of winter. This variety was first noticed in a field of normal-sized tobacco plants around 1906 because it grew to the great height of about 2.5 m and produced as many as a hundred leaves for harvest. Commercially, this was a huge advance over other varieties.

Garner and Allard tried growing Maryland Mammoth under many different conditions before discovering why flowering began so late in the season.

DAY AND NIGHT LENGTH: PHOTOPERIODISM

Flowering in Maryland Mammoth explained
Garner and Allard concluded that:

- at the most critical time of year for flowering in Maryland Mammoth the days were too long and the nights too short at the latitude of Washington, DC;
- their plants did not begin forming flower buds until the length of day had shortened to between 10 and 12 hours;
- in daylight hours longer than this the plants went on forming new leaves indefinitely;
- in the Washington area, the days do not become as short as 10–12 hours until late summer. By then it was too late for this tobacco variety to flower *and* produce mature seed before winter.

Categories of photoperiodic response
Garner and Allard went on to test the discovery of a link between flowering and day length with many other species. They found that plants could be placed in three major categories:

- those needing days of less than 14 hours to set flowers (called **short-day** plants);
- those needing day length longer than 14 hours, but not continuous light (**long-day** plants);
- and those not choosy about day length (**day-neutral** plants);

Garner coined the word **photoperiodism** to describe these responses.

Examples of responses
Many more examples have now been added to the categories suggested by Garner and Allard.

Examples of short-day plants are annuals originating in the tropics where short days are normal, year round, and most species found in temperate areas of the globe which flower either in spring or in autumn when day length is shorter than in mid-summer, late-blooming chrysanthemum being one of the best examples.

The long-day plant category includes wheat, potato, lettuce, beet, and other species of temperate regions which flower in the long days of mid-summer.

Among the day-neutrals are plants like maize, tomato, most weeds, and the many tropical plants which produce flowers and leaves year round, having in fact no specific annual rhythms.

PHOTOPERIODISM EXPLAINS MUCH

From Tournois's and Garner and Allard's astute observations followed by other, more recent, studies, it is now clear why certain plants flourish in some latitudes and are absent from others.

In the tropics, where day length remains close to 12 hours, only short-day and day-neutral plants succeed in flowering. Long-day plants do not flower where day lengths never exceed about 13 hours.

In very high latitudes, such as the Arctic, only long-day and day-neutral plants are found. Short-day plants do not flourish there because, although short days eventually come, they do so only briefly and too late to allow flowering and seed formation before sub-zero weather descends.

Synchronized flowering

Garner and Allard's explanation of flowering based on photoperiodism also provided a reasonable explanation of why all individuals of a certain variety flower at about the same time in a particular location even though planted at different times during a growing season.

Exposure to the correct day length for a few days or weeks is usually necessary to trigger flowering no matter when a species is planted. If short-day plants are kept indefinitely in long days they will never flower. Similarly, long-day plants kept indefinitely in short days also continue with leafy growth, never forming flowers.

But then, after a few weeks of growth under the wrong regime for flowering, if the two groups of plants are switched so that they are now in correct day and night lengths, production of flower buds goes ahead in all individuals, at the same time, no matter whether they were planted at the same time or not.

Commercial exploitation of photoperiodism

The discovery that the flowering of plants could be controlled by artificially setting day length quickly led to major advances in the marketing of some major horticultural crops. One of the first to be exploited was the chrysanthemum which, in the 1920s, was a fall-blooming plant in northern latitudes. But Allard was able to advance flowering of this species from mid-October to mid-July by darkening the greenhouses where the plants were growing. He planted chrysanthemums indoors in the spring. As the days grew longer in May and June, he darkened the rooms housing his plants to keep day lengths short, advancing the date of flowering by about 3 months.

In the 1930s, horticulturists began following Allard's lead and found they could produce chrysanthemum blooms at will at any time of year. To keep cuttings in leafy growth until flowering was desired they took to extending the short days of winter as well by using artificial light in their greenhouses so that chrysanthemums could be produced for the Christmas market. Such artificial manipulation of day length is now routine for many horticultural crops.

HOW PLANTS "SEE" LIGHT

Once the idea had taken hold that day length could control flowering in plants, there arose a curiosity about how a plant could possibly "see" light. Also, there was a growing amount of information about how light and dark influenced biological clocks in plants (see Chapter 8). But how do plants "see" light and convert that information into actions?

Attempts aimed at answering this question have provided brilliant insights into plant growth and development until today an entire new branch of science has blossomed. It is now realized that many aspects of plant growth are under the influence of light, not just flowering.

NIGHT LENGTH VERSUS DAY LENGTH

Effect on short-day plants

A critical breakthrough in understanding how plants perceive light was made in the late 1930s. Using cocklebur (*Xanthium strumarium* L.), a short-day plant, it was discovered that the trigger for flowering was not linked to light but to dark. It became clear that "short-day" plants should really be called "**long-night**" plants. If a plant such as cocklebur was kept in short days (8 hours light, 16 hours dark), it flowered. If, however, the 16-hour night was interrupted mid-way through by exposure to light, the plants would not flower. The exposure to light during the night did not have to be very long; a few minutes was enough. Nor did it have to be very bright; a little brighter than moonlight would do.

Here, then, was a puzzle. Very short periods of light in the middle of the night seemed to signal the plant that it was not in short days and long nights but long days and short nights. A long, uninterrupted span of darkness seemed essential to sending the signal to the short-day plant to flower.

Effect on long-day plants

The opposite of a light interruption of the long night favored by short-day plants had no effect on long-day plants. If a plant that required long days and short nights for flowering had a short dark period imposed on it in the middle of its long day, flowering still proceeded. However, keeping a long-day plant on short days while giving a few minutes of light mid-way through the long night caused flowering. Just as short-day plants might be renamed "long-night" plants, so long-day plants might be called "**short-night**" plants instead.

Tournois's results confirmed

The conclusion drawn by Tournois in his experiments on Japanese hop was, thus, fully confirmed. Night length, not day length, is the decisive factor in determining whether or not long- and short-day plants will flower. The importance of this early observation was increased when it was realized that knowledge of the effects of small amounts of light on plants in the middle of their night periods (as Stoppel had first found with bean plants: see Chapter 8) was the key to answering the question of how plants "see" light.

AN ANSWER TO HOW PLANTS PERCEIVE LIGHT

Importance of wavelength

An obvious next step in investigating the effect on flowering of short periods of light during the night was to ask whether any particular wavelength was more effective than others in promoting or preventing flowering. An answer again came from the use of the cocklebur plant where it was discovered that flowering could be prevented if short periods of **red** light were given mid-way through a long night. Long-day plants, similarly, could be made to flower if short bursts of red light were given mid-way through their long nights.

Such results were at first mystifying. Why red light? The answer came not from further investigation of flowering but in other ways.

EFFECT OF LIGHT ON SEED GERMINATION

Small seeds and light

In the nineteenth century it was discovered that exposure to light improves germination in plants with small seeds. If a small seed germinates deep in the soil, it does not have sufficient food reserves to sustain the growth of its seedling until it reaches the surface of the soil and begins photosynthesizing.

Small seeds tend not to germinate until some activity around them carries them close to the soil surface. An animal (a rodent, an earthworm) burrowing into the soil, turning it over in the process, causes seeds to be thrown up onto the soil surface. Cultivation of soil by gardeners or farmers serves the same purpose.

Weed seeds and light

Many seeds respond in this way, which is why newly cultivated soil seems to become "weedy" so quickly. Weed seeds may lie dormant deep in the soil for years: "one year's seeds, seven years' weeds," is an old gardeners' saying. Weeds are past masters at spreading germination over many years as and when their seeds are turned up to the soil surface during cultivation. Up to a billion weed seeds per hectare of cultivated land may await the light needed to release them from their dark-induced slumber.

Lewis H. Flint and Edward D. McAlister: lettuce seed

Lettuce (not all varieties, but several) is one type of cultivated seed that requires light for germination. Lettuce has small seeds that remain dormant in the soil unless exposed to light after they are planted and have become swollen with moisture (dry seeds do not respond to light).

In 1934, **Lewis H. Flint** was hired by the United States Department of Agriculture and given the task of finding what kind of light most effectively caused light-requiring lettuce seed to germinate. He soon discovered that just 4 seconds of sunlight prompted swollen Arlington Fancy lettuce seed to sprout.

EFFECT OF RED AND FAR-RED LIGHT ...

...On lettuce seed germination

Flint was joined in his work by the physicist, **Edward D. McAlister**.

- Together, they found that germination of lettuce seed was increased to the greatest extent by red light, just as had been found with flowering

- They also found that far-red light prevented it
- If they treated moistened seeds with red light, germination occurred
- If red light treatment was followed immediately by far-red, the effect of the red light treatment was canceled.

Seeds seemed to have a memory, reacting to whichever light treatment was given last.

...On flowering

An obvious question then was whether red and far-red light had a similar effect on flowering?

When short bursts of red light were given to the short-day plant, cocklebur, in the middle of a long night, followed at once by far-red light, the plants ignored the fact that they had been exposed to red light, and flowered. It was as though the plants had been exposed to no light at all.

Red light given to long-day plants in the middle of a long night would normally signal to the plant that it was in a short night. This signal was also reversed by far-red light given right after red light treatment.

A THIRD EFFECT OF RED AND FAR-RED LIGHT

Harry Borthwick and Sterling Hendricks

These observations on seed germination and flowering stimulated the curiosity of two other investigators, **Harry Borthwick** and **Sterling Hendricks**, who began studies which were to lead to a discovery that continues to inform and enrich our understanding of plant growth and development.

Between about 1945 and 1960 in the USA, Borthwick and Hendricks investigated the effects of red and far-red light on several aspects of plant growth and development, not just one as previous investigators had done. They repeated and extended the work of Flint and McAlister on seed germination; they did the same with the work of Garner and Allard and others on the influence of red and far-red light on flowering. They, themselves, investigated a third effect of red light on plants.

Etiolation

In many kinds of plants, when seeds are germinated in the dark, the young seedlings have long, spindly stems and tiny leaves which are yellow, not green, a condition called **etiolation**. Most plants grow in a "leggy" fashion if not given enough light; seeds germinated in complete darkness show this condition in the extreme.

Photomorphogenesis

Exposure of such sickly, weak-looking seedlings to a few minutes of sunlight causes a rapid transformation to take place even if the seedlings are returned to darkness.

- The most noticeable change is in the leaves, which begin to turn green and to grow to normal size within a few minutes of exposure to sunlight
- The spindly stem slows its growth in length, becoming thicker and sturdier
- Many other, less obvious, changes in the growth form of the plant occur at the same time, a process called **photomorphogenesis**.

The rapid, spindly growth of seedlings which typifies etiolation is an adaptation many plants have evolved. Many seeds, other than those that are light-sensitive, normally germinate in the dark deeply buried in the soil. The smartest thing for a seedling in this position to do is to grow as quickly as possible to the soil surface so that photosynthesis can begin. Until it reaches the light, a plant does not need leaves, so no leaves of any size are formed, just tiny vestiges. The food stored in the seed is used to extend the stem as quickly as possible through the soil to the light.

Red light, far-red light, and etiolation

As might be suspected, red light also causes the leaves on etiolated seedlings to turn green as well as grow to normal size, spindly stems to slow their growth and become normal in appearance. And again, as might be suspected, far-red light reverses the effect of red light: etiolated plants continue their rapid, spindly growth if a red light treatment is followed quickly by far-red light.

The important insight Borthwick and Hendricks drew from these observations on the effects of red and far-red light on seed germination, flowering, and etiolation was that they must be linked, a conclusion which led them to a brilliant, intuitive leap of imagination.

AN INTUITIVE LEAP

Artificial conditions

Borthwick and Hendricks hypothesized:

- that there was in plants a single compound, a pigment, capable of absorbing red and far-red light;
- they suggested that the pigment was converted to an active form by red light, and to an inactive form by far-red light;
- they also hypothesized that any pigment previously converted to an active form by exposure to light reverted to the inactive form in the dark.

Natural conditions

Since there is more red than far-red light in the Sun's spectrum at the Earth's surface, exposure of the pigment to sunlight would convert a large fraction of it to the active form, information used by the plant to sense the presence of light. A plant could then use the information to trigger a variety of responses.

- Simple exposure of seeds to light at the soil surface causes them to germinate
- In short- and long-day plants, red light could act as a Zeitgeber to determine the length of day and night by sensing the appearance of light at dawn or its disappearance at dusk, linked to the seasons, information used to sense whether a night was long enough or short enough to trigger flowering
- In etiolated seedlings growing through soil, red light activation could be used as the trigger to "tell" the plant when daylight had been reached and normal growth could begin
- Under natural conditions, the active form of the pigment, which is unstable, reverts to the more stable, inactive form at night.

PHYTOCHROME: THE BLUE PIGMENT

Their intuitive leap led Borthwick and Hendricks by the early 1960s to the isolation from plants of a single blue-colored pigment (blue because it absorbs red and far-red light), later called **phytochrome**. All flowering plants have it in every living tissue. Conifers, mosses, ferns, and at least some, if not all, algae, even some bacteria, also contain phytochrome or a pigment similar to it.

The list of processes in plants under the control of phytochrome is growing all the time as we learn more about its effects. All the processes covered already in this chapter and the one before are known to be under phytochrome control.

DISCOVERY OF PHYTOCHROME CRUCIAL TO UNDERSTANDING PLANT GROWTH AND DEVELOPMENT

The discovery of phytochrome was a crucial step in understanding how plants control their growth and development as a whole. The lives of plants are dictated by light.

- They exploit light as a source of energy to produce food in photosynthesis
- Stem tips sense the direction of light and use the information to orient shoots towards the light (see Chapter 7)
- Information about length of days and nights and the seasons is also needed by plants so that their cycles of growth activity can unfold in an orderly way
- For seed germination, vegetative, and reproductive growth, the need for information about light and dark conditions is clear
- To these needs should be added others such as the requirement for information about the approach of winter weather. The plant "knows" winter is not far away because the days are becoming shorter, nights longer. Thus, the plant can produce winter buds in mid-summer (some trees, for example, do so in August in the northern hemisphere), or at any other time.

The negative influence of phytochrome

So far, the positive effects of light on plant growth and development have been emphasized, but a negative influence can also be used to advantage.

In desert species, exposure to light indicates to a seed that it is at the surface of the ground and should *not* germinate because of lack of water there or, perhaps, because of excessive heat. Conversely, a seed buried deep underground could sense the *absence* of light through the lack of the active form of phytochrome, information it could use to signal that it is safe to germinate since well below the soil surface moisture is more likely to be found.

A simple, useful analogy

A simple analogy might help to illustrate how just one signal might be linked to so many events all at the same time.

When an electrical switch is thrown an electric current is either switched on or off. What action then occurs depends on to what the wire carrying the current is connected. It may be linked to a light bulb, an electrical heater, a fan, an alarm, or any number of other appliances. The switch is just a trigger which can activate or deactivate any number of processes depending on connections.

So it is with phytochrome. In all cases the "switch" is the same. What switching phytochrome "on" or "off" triggers depends on its connections. If the pigment is in a seed it may be connected to germination; if in a leafy branch, it may be linked to flowering; if in an etiolated seedling, it may be a trigger connected to normal growth.

LIGHT AND PLANT DEVELOPMENT

That light is so often at the center of the regulation of plant development is not surprising given the path of evolution within the plant kingdom. Nor is it surprising, therefore, that the discovery of phytochrome was just one step along the path of discovery of an array of pigments which plants have evolved to sense light.

In Box 9, a brief account is provided to illustrate the sophisticated photosensory systems plants have evolved for acquiring and interpreting information contained in light.

BOX 9. SEEING THE LIGHT

Plants have four classes of pigments for intercepting, interpreting, and using the information conveyed by light to regulate their development.

__Phytochrome__ exists in two states: a red form (Pr), which absorbs red light maximally at 665 nm, and a far-red form (Pfr), which has maximum absorption at 730 nm. The two forms are photoreversible: light of 665 nm converts Pr to Pfr, which is the physiologically active form; absorption of light at 730 nm by Pfr converts it back to Pr, the inactive form. Pfr is also unstable and will revert to Pr in the dark.

Most angiosperms have at least three distinct phytochromes (phyA, B, and C), although five have been found in Arabidopsis *(phyA–E). The members of the phytochrome family are expressed at different times in different tissues, and in different combinations during a plant's life cycle. These variations allow them to mediate different responses to light and to interact with other photoreceptors in complex pathways of control, by no means yet fully understood, of plant development (Figure 9).*

The first blue light receptor was isolated from Arabidopsis *in 1993, recognized because of its absence from mutants which would no longer respond to blue light; it was given the name __cryptochrome 1__ (cry1). A second, __cryptochrome 2__ (cry2), has now been isolated; its roles in plant development overlap with those of cry1; both absorb in the same range of wavelengths of blue (400–450 nm) and UV-A (320–400 nm) light.*

Blue and UV-A light regulate many growth and development phenomena in plants, animals, and fungi. In plants, stem elongation, opening and closing of stomata, timing of flowering, and the setting of biological clocks are some aspects of development affected by blue and UV-A light.

During his investigations of the effects of light on plants, Charles Darwin noted that the phototropic movement of

BOX 9. (cont.)

Phytochrome gene family	Some functions in *Arabidopsis*
phyA	Seed germination Hypocotyl elongation
phyB	Hypocotyl elongation Flowering Seed germination
phyC	Hypocotyl elongation
phyD	Petiole elongation
phyE	Seed germination Petiole elongation

FIGURE 9 The dicot, *Arabidopsis*, has five phytochrome gene families (phyA–phyE). Sometimes, more than one has a role in a single growth phenomenon. For example, phyA, B, and E all have influence on seed germination. Not all dicots have all five families and their roles are not necessarily the same as in *Arabidopsis*. Monocots seem to have only three families: phyA, B, and C (after Taiz and Zeiger, 2006).

coleoptiles towards light was lessened if the light was passed through a solution of potassium dichromate, which strongly absorbs blue light. It is now known that a pigment, **phototropin**, mediates phototropic responses over the same range of wavelengths as cryptochromes have their influence on plant development.

The three classes of photoreceptors above are collectively referred to as **chromoproteins**, all of which contain a component for absorbing light, the **chromophore**, attached to a protein component, the **apoprotein**. The members of each class differ from one another genetically in their protein components, which are products of different genes.

A fourth class of pigment, not yet characterized, exists in plants, which mediates responses to low levels of UV-B light.

SUMMARY

- Plants use light to regulate and direct major milestones in their growth and development
- Many species, which flower at different times of year, use the diurnal cycling of light and dark to trigger flowering or, conversely, to prevent it occurring until the right conditions are sensed, a process called photoperiodism
- The right conditions may be when day length is shorter (e.g. short-day plants flower in spring or fall) or when it is longer (e.g. long-day plants flower in the summer months)
- Other species are indifferent to light (e.g. day-neutral plants); they flower over a wide range of day lengths
- Night length rather than day length seems to be particularly important in directing photoperiodism
- When seedlings first emerge from seeds deep in the soil, they take on a growth form which takes them to the soil surface as quickly as possible so that photosynthesis can begin. This spindly growth form is called etiolation and also occurs in mature plants if they find themselves in unfavorably deep shade for extended time periods
- When exposed to light at the soil surface, etiolated seedlings begin to take on a normal growth form, a transition called photomorphogenesis
- Some seeds will not germinate at all until exposed to light. They may remain dormant in the darkness of the soil for many years (see Chapter 10)
- These various responses to light and dark are sensed and controlled by a single plant pigment, phytochrome, which is sensitive to red and far-red light
- Phytochrome is just one of a group of pigments which govern the orderly progression of a plant through the phases of growth and development over its lifetime (see Box 9).

10 Dormancy: a matter of survival

INTRODUCTION

In preparation for the most unfavorable weather, a plant may need special protection against the climate. These periods of recurring poor growing conditions must be anticipated well in advance. It would be no use for the plant to begin preparing for winter, for example, the morning of the first frost or for a long, dry, hot season in a desert when water was no longer available.

What a plant does in preparation for long periods of poor weather is often quite elaborate, requiring a long period of good weather *after* the signal is received that an unfavorable season is approaching. The signal received by the plant cannot be linked directly to future poor conditions. For example, it is *not* low temperature which triggers the processes inside a plant leading to preparations for winter. Preparations might begin in mid to late summer when the temperature is still high. Shortening day length is a more reliable signal than temperature for plants to use to anticipate winter.

SURVIVAL STRATEGIES

Winter buds

When forming winter buds a plant stops producing new leaves. Instead, small, tough scales are formed which tightly enclose the soft, delicate growing points in terminal buds on branches. They can withstand freezing and thawing many times over without disintegrating and they repel water while keeping the tender tissues inside moist and alive. Only in the spring, when their task is complete, are they shed as the growing points begin once more to grow.

Storage organs

Many plants die back in winter or during a dry season, retreating beneath the soil surface for protection. Stem tubers (potatoes, for example) are regions of the stem which become swollen while other parts of the plant die back. Bulbs (tulip), corms (hyacinth), tap roots (carrots), and root tubers (dahlia), are examples of the various ways plants achieve the same thing, production of storage organs to withstand long periods of adverse conditions. During the preceding growing season, food is delivered to these organs for use the next growing season.

Seed-bearing plants

Above all, important and successful as these strategies are, the development of the seed as a means of survival is the most widespread and important; seed plants have inherited the land.

Two hundred million years ago, the Earth was dominated by the "spore bearers," plants such as club mosses and horsetails which often grew in great profusion. Some tree-like horsetails grew 30 metres high and had massive trunks rather like those of the modern palm. The ancient club mosses, horsetails, and another, now extinct, group, the *Calamites*, dominated the Earth's vegetation. But, as later with the dinosaurs, rapid climatic changes found these giants unable to adapt. Their massive corpses gave rise to rich fossil deposits in the shales, the origin of our coal deposits. They survive today as smaller plants, often in the under-canopy of forests along with the ferns which were also spore bearers more dominant than they now are.

So, what was it about the seed bearers that made them so much more successful than spore bearers?

Genetic diversity through the seed

The seed is a product of sexual reproduction, the fertilization of an egg by pollen. Just as in animals, sexual reproduction is the basis of genetic diversity as well as being the bridge from one generation to the next. Many types of spores can be produced without sexual reproduction; offspring produced this way are identical to the parent. A plant produced from a seed is different in some way from its parents.

Lack of genetic diversity puts spore bearers at a disadvantage during times of change in climate; climates across our planet have changed many times through the eons and will continue to do so. Genetic flexibility is a great advantage at times of change. If offspring resemble parents too closely, and if the parents are ill-adapted to change, then their offspring will be also. Variability in offspring greatly increases the chances that at least some will survive even if those that too closely resemble their parents do not.

Seed dispersal

The seed is also a way for a plant to disperse itself in a protected way. The embryo inside the seed is surrounded by a store of food to nurture it in the hours or days after germination begins. The tough outer shell or coat of the seed protects the embryo from adverse weather as well as from attack by disease organisms and predators. Dried seeds are just about indestructible. Many seeds, when just formed and still full of moisture, can be killed by exposure to temperatures around the freezing mark. The same seeds, when fully mature and dry, can withstand temperatures of $-80°C$ or even lower. Durability is the hallmark of the seed, with a few exceptions.

The viability of seeds

The longest authenticated viability known is among the 2000-year-old date palm seeds retrieved from the Masada fortress at the edge of the Dead Sea; second in line is a 1300-year-old Chinese lotus seed. There are a few other credible estimates that put the ages of some viable seeds at between 400 and 600 years; much more typical lifespans are, however, from 10 to 50 years.

Not that seeds are always long-lived. Those of silver maple, wild rice, and Japanese willow lose viability in a week in air.

Others live for only a few weeks or months, dying if they lose even a small amount of moisture or are exposed to lower than normal temperatures (e.g. seeds of tropical trees). The seeds of major crop plants germinate a short time after planting although their wild ancestors may have had the ability to delay theirs. Domesticated varieties of food plants have been bred over many centuries for quick germination. Farmers do not want any delay in seed germination; they determine for themselves when to plant or not plant their crops.

DORMANCY

Of interest here is how long-lived seeds remain viable for so long, a condition referred to as **dormancy**. But what are the causes of dormancy? What advantages does dormancy have? How do seeds break out of their dormant state and germinate? What is dormancy in the first place?

We tend to use the word dormancy or dormant in a broad way to mean any state of suspended activity. Volcanoes are dormant between eruptions, as are hibernating animals between periods of activity. Insects are often said to be dormant before they emerge from their pupal stage. But in plants the word should be used with greater care.

True dormancy in plants

True dormancy is a special property with its own set of conditions. Some distinguish between true dormancy and what is called **quiescence**, which is just the prevention of growth by the absence of one of the basic conditions needed for growth, such as insufficient water or too low temperature. Supply the missing ingredient and the plant resumes growth. This is not so with true dormancy. Even under ideal growing conditions, dormant seeds, winter buds, or storage organs will not resume growth at once. The state of suspended animation we call dormancy used by plants to avoid climatic stresses can often be very deep, not easily reversible.

Advantages of dormancy

The ability to avoid climatic stress has been of enormous importance in the evolution of plants, allowing them to colonize places that are hostile to growth for periods up to many years.

Most plants have adopted a non-mobile lifestyle which is made possible because they make their own food by photosynthesis from simple substances. However, when it is time to disperse themselves, they need to be mobile, which means detaching a portion of themselves in a form that can withstand possibly long periods of adversity. This requirement is met by the production of seeds, whose germination is under controls which are responsive to environmental conditions.

Seed dormancy spreads out germination

For a seed to germinate it needs water, oxygen, and the right temperature. Some germinate as soon as they are ripe but many do not even when conditions are ideal. This dormant period may simply be to ensure that a seed germinates in the spring following the year it was produced. Spring usually provides better growth conditions than does autumn when a following cold winter might destroy tender seedlings (Figure 10). The seeds of some Mediterranean plants are prevented from germinating by high temperatures to avoid scorching summer conditions.

In many other cases, dormancy helps to ensure that at least some members of the batch of seeds produced by a plant in any one year germinate each succeeding year over a long period. In some plants, the clovers being an example, a proportion of seeds is programmed to germinate at once while others have a delay mechanism. The weed, Fat Hen, is particularly elaborate, having one kind of seed which is large and germinates immediately and three other, smaller kinds which germinate at various intervals. The fruit of cocklebur has two kinds of seeds, one dormant, the other, not. The latter germinates as soon as it leaves the plant; the former remains in the soil until the seed coat is damaged in some way.

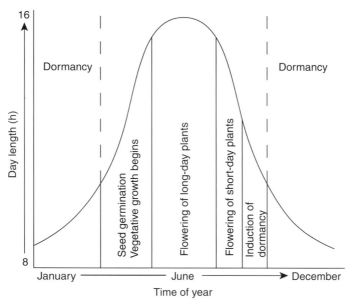

FIGURE 10 How day length influences key developmental steps in plants found in temperate zones (after Raven *et al.*, 1999).

TYPES OF DORMANCY

IMPOSED DORMANCY

One major characteristic of dormant organs is their ability to tolerate low water content. As much as 90% of the weight of a growing plant is water. In contrast, many seeds have 10% water or less, but not zero. Many seeds *must* dry out in this way before they are able to germinate (a period of **after-ripening**). This is an imposed dormancy mechanism typical of plants which grow in places with hot, dry seasons where seeds that germinate immediately would have great difficulty in establishing themselves.

SEED COAT DORMANCY

One of the simplest and most effective types of dormancy is that imposed by the seed coat. Seeds made dormant in this way are often referred to by farmers as being "hard" (clover is one good example). Water is prevented from entering hard seeds by a thick, waxy surface which has to be ruptured before germination can begin. A breach of the seed coat barrier can be achieved in many ways, each of which is subject to many variables so that some seeds may have their defenses breached quickly; others may lie in the soil for months or years before water finds a way to penetrate. In this way plants can spread out their seed germination over long periods, which increases the chances of at least some germinating successfully.

Seed coat dormancy broken ... by fire

A grass or forest fire may weaken seed coats in large numbers and allow rapid and widespread germination among hard-seeded plants. In one case, an *Albizzia*, a small legume tree found in Western Australia, the seed coat has in it a tiny corky plug; heat from a fire may cause the plug to pop out, allowing water and air to enter. Seeds in conditions such as those of the chaparral of Mediterranean climates are breached by fires that routinely engulf these areas. The frequent fires that sweep through Southern California, fanned by the Santa Ana winds, are one good example. The result is a rapid recovery of an area following a fire.

... by scarification

Soil movement caused by ice formation and melting may cause abrasion (**scarification**) of seed coats, which become permeable to water. In farming, the coats of hard seeds are deliberately scarified, mechanically, just before they are planted so that water can enter at once.

... by other organisms

Passage through the digestive tracts of animals, notably birds, is another way to weaken a seed coat. In one bizarre example, a particular tomato, which grows in the Galapagos Islands, produces seeds which must pass through the digestive tract of the giant tortoise before they will germinate. Presumably, this slow, lengthy transit is needed to complete the process of weakening this particular seed coat.

Attack by insects and other soil organisms as well as fungal and bacterial diseases may also weaken seed coats enough to allow germination to occur.

EMBRYO DORMANCY

In other cases, the seed coat is no barrier to the entry of water and yet the seed still will not germinate. Some embryos, such as hogweed, are not fully formed when the seed is shed and will not grow until their development is completed. In others, like the European ash, embryos are complete but have not accumulated all their food reserves at the time the seed is shed. This type of dormancy is often associated with the need for a certain temperature range before germination will occur. For instance, hogweed embryo development takes place best at just 2°C. The seed can then mature over winter and early spring before germinating when the soil warms the following year.

DORMANCY LINKED TO ...

... *light and dark*

Light is essential for germination of seeds like foxglove, tobacco, many primulas and some lettuces (see Chapters 8 and 9). Many plants show a kind of seed dormancy linked to light, especially when the seeds are first formed. As they age in storage some seeds lose their responsiveness to light; others become more sensitive to light if attempts are made to germinate them at unfavorable temperatures.

Responses to light are of every kind imaginable; sometimes germination is triggered by simple exposure to light; other seeds are inhibited by light; yet others need day lengths of a certain kind, either long or short days. Germination of foxgloves in woodland clearings is often spectacular, indicating that the seeds are capable of discerning not just the presence or absence of light but variations in its strength. This photocell-like capability is used to measure light quality so that germination occurs only when the seeds are in bright, near full, sunlight.

Of course, the seeds are sensitive to light only when imbibed with water, not when dry.

Other seeds, such as those of many desert plants, germinate only in complete darkness, deep in the soil, as long as there is adequate moisture. Such a response ensures that a seedling has a chance to root before being subjected to searing desert heat. Seeds of this sort can have their dormancy "reset" by even a brief few minutes of exposure to light.

... low temperature and oxygen exposure of seeds

The seeds of stone fruits (peach, cherry, plum), some deciduous trees, and several conifers will not germinate until exposed for weeks or months to low temperatures and oxygen levels under moist conditions. Rarely, the opposite (high temperature) is essential; and some will respond only to alternating high and low temperatures.

The need for a period of low temperature before germination will occur protects the seeds of plants in temperate regions from premature germination in the fall or during an unseasonally warm period in winter and is by far the most prevalent form of dormancy control by temperature. In some seeds, germination is merely speeded up; others remain dormant indefinitely unless exposed to temperatures close to 5°C.

... low temperature exposure of some whole plants and plant organs

A similar requirement occurs in the case of certain whole plants or plant organs (bulbs, storage roots, winter buds, etc.) which enter a state of winter dormancy that can only be broken after a period of **prechilling**. For example, if peach or apple embryos are removed from unchilled seeds and germinated, stem elongation is slow and irregular, leaves distorted. These symptoms can be relieved by chilling the stunted seedlings at 5°C, a condition that in natural surroundings would be encountered sooner or later.

... the seasons

Temperature linked to the seasons controls germination of many desert plants. Thus, in the Colorado desert of the United States, if it rains and the temperature is only 10°C, it is mainly the winter annuals that germinate; between 26°C and 30°C, summer annuals emerge; cacti germinate best between 30°C and 40°C. Seedlings in these regions also sprout very rapidly to ensure their establishment before moisture runs out.

CONTROL OF DORMANCY

It is easy to understand failure to germinate in seeds where a mechanical barrier exists. In seeds with hard coats that have to be breached, there is a visible reason why germination is delayed. The same is true of those seeds which contain immature embryos.

But what prevents germination in other cases; for example, when the seeds are complete yet will not germinate, sometimes for years, and in winter buds on deciduous trees or in storage organs like tubers, bulbs, and corms? What happens during prechilling? Why is light (or darkness) required? In some cases, seeds inside succulent fruits will not germinate; for instance, the temperature inside a tomato is usually ideal for germination and there is ample moisture and oxygen, yet seeds will not grow out. Why not?

Chemical inhibitors

The answer is straightforward in the case of the tomato: seeds germinate if removed from the fruit, indicating that there is nothing about the tomato seed itself which is lacking; it is the juice of the tomato fruit which prevents germination. The seeds will germinate only after the fruit flesh has either decayed or is eaten by an animal, a seed dispersal mechanism. Many fleshy fruits inhibit their seeds in this way.

In other cases of fruits, seeds, dormant buds and storage organs, chemical inhibitors within the dormant organ itself must be destroyed before germination will proceed. Sometimes, it is a particular growth substance which arrests growth. Over time, the inhibitor is broken down and its influence over germination removed (see Box 10).

Other chemical inhibitors present in seeds must be leached out before germination can proceed; in nature, enough rainfall to dissolve out these substances also leaves the ground wet enough to allow seedling survival. This is particularly important in deserts where lack of moisture is more of a limit to growth than any other factor. In desert plants, brief showers will often have no influence on seed germination; only relatively heavy rainfalls will provide enough moisture to complete leaching and promote germination.

Seeds, certainly, and maybe other kinds of dormant organs too, contain many kinds of inhibitors other than growth substances. The inhibitor may be as simple as table salt which, in the case of shadscale, is in such high concentration in the seed that it prevents germination until leached out. However, usually the inhibitor is more complex than salt. Some seeds release cyanide, especially members of the rose family; others, ammonia; yet others contain mustard oils, common in the brassicas, or alkaloids and steroids, or many other chemicals which we use as medicines. Plants have a wide range of chemicals that they use both to control their own development, including germination of their seeds, and, as we shall see in later chapters, to defend themselves against diseases, predators, and competition from other plants.

SUMMARY

Dormancy in seeds and other organs is one more component of the array of survival strategies plants employ.

- In some cases, this strategy takes the form of reduction in water content in seeds while they are still on the parent plant. These seeds remain dormant for a minimum period even if they are dispersed while there is still sufficient moisture for germination. Seeds of winter annuals are examples of this type of dormancy
- When first produced, the seeds of many wild annuals will not germinate under normally favorable conditions, but only after encountering a

period of cool temperatures during their dried down state. As the period
of dryness progresses they pass from a state of deep dormancy to one
of relative dormancy, during which they become increasingly able to
germinate when temperatures are favorable

- Difficulty arises in desert areas subject to infrequent and irregular rains.
A light fall of rain might allow germination of those seeds that have
completed their dormancy period but might not be enough to support the
growth of seedlings; germination occurs only after heavier rainfall. It is
thought that light rains are insufficient to wash out the inhibitors of germination lodged in seed coats. Continued rainfall for a period leaches out
the inhibitors sufficiently to allow germination and also provides sufficient
water to soak the soil for subsequent growth, flowering and seed set

- The requirement for chilling of seeds and buds is characteristic of plants
inhabiting temperate zones where cold winters are experienced. This
requirement ensures that seeds and buds do not begin growing until
winter is past

- Light sensitivity is of value to dormant organs not just for sensing day or
night in the environment but also for monitoring light quality. For example, a canopy of leaves overhead alters the wavelengths of light available by acting as a filter. When light passes through the canopy, some
of the red and blue wavelengths are absorbed by leaves. Since red light
is required by some seeds for germination, the absence of red light can
act as a signal to those seeds to remain dormant. Later, perhaps when
the leaf canopy has thinned or has opened up for one reason or another,
more red light will penetrate to the under-canopy and germination of
light-requiring seeds will proceed. This may therefore be responsible for
germination control in deciduous forests, or even under herbaceous plant
cover, where many seeds germinate in early spring before leaves form a
canopy, and also for seedlings growing in forest clearings

- The many mechanisms of dormancy emphasize that the period between
the production of a seed and its germination is not just about dispersal and
finding a suitable location in which to sprout. The period may be drawn out
for reasons of survival. In most cases these time lapses ensure germination
either at some propitious moment or as opportunities arise over time

- Prominent among the chemical inhibitors of dormancy in a wide variety
of seeds is the plant growth substance, abscisic acid (ABA). It is only
after the concentration of ABA in a seed falls to a certain threshold,
establishing a ratio between the inhibitor and other hormones favorable
for growth, that germination proceeds (see Box 10).

BOX 10. SEED DORMANCY: EVOLUTION AND
CONTROL

*It can only be speculated as to what selection pressure might
have been placed on seeds to bring about the evolution of dormancy. A major influence might have been whether or not seeds
accumulated in the soil, i.e. formed soil seed banks. Such banks
will only become established if seeds are capable of combating
such hazards as competition between individuals or predation
by animals, insects, and microbes, to name but two strong influences on survival. Hazards such as these could be mitigated by
spreading seed germination over a longer period of time rather
than having it occur in all seeds immediately after their release
from the parent plant.*

*It is also a matter of speculation as to which form of
dormancy was the first to evolve. That due to underdeveloped
embryos at the time of seed release is thought to be, possibly,
the earliest, followed by evolution of seed coats impermeable
to water, then, water-impermeable coats combined with some
form of physiological dormancy (found in angiosperms only, not
gymnosperms) and, finally, physiological dormancy, the most
widespread today.*

*Prominent among chemical inhibitors is the plant growth
substance abscisic acid (ABA), which has long been known to
play a key role in physiological seed (and bud) dormancy in many
angiosperms (see also Chapter 7). ABA produced by the embryo
is particularly crucial. The hormone begins to increase early
in embryogenesis, peaking during mid to late stages of embryo
development. During these later stages of development, seeds
accumulate and store the compounds (e.g. carbohydrates, lipids,
or proteins) needed to support embryo and seedling growth at
germination. At this stage, they also produce in their embryos
the proteins needed to protect against injury as seeds dry out.*

BOX 10. (cont.)

The balance between hormones is of central importance in establishment of and release from physiological dormancy in seeds. For instance, if the ratio of ABA to gibberellic acid (GA) favors the former, dormancy deepens; as, over a variable period of time, the ratio changes to favor the latter, GA-induced synthesis of the many enzymes essential for the breakdown of stored seed reserves increases, allowing germination to proceed. For example, GA stimulates the aleurone layer of cereal grains to begin forming α-amylase, an enzyme which aids in the breakdown of starch; high ABA levels inhibit α-amylase production.

Recent studies of seed germination have identified interactions between ABA and two other hormones, ethylene and brassinosteroid. Taken together, investigations of the involvement of plant growth substances show that there is a complex regulatory web of interactions between several hormones and the steps leading to growth of the next generation.

No matter what the method of dormancy evolved by a species of plant, survival to the next season or to the next generation is the objective.

11 Color, fragrance, and flavor

INTRODUCTION

Plants produce a bewildering array of exotic chemicals: the pigments which give color to leaves, petals, fruits, and seeds; the substances that create aromas and tastes; those formed to help defend against attack by diseases, predators, and competitors; and others with no known function (Figure 11).

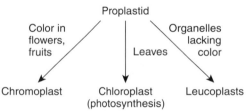

FIGURE 11 Plants produce a family of plastids all derived from the proplastid. The most familiar is the green chloroplast. Less familiar is the other colored plastid, the chromoplast, which contains the pigments giving color to flowers, fruits, etc. Least conspicuous are the colorless leucoplasts, which may act in storage (e.g. amyloplasts store starch) or as organelles where a wide array of chemicals may be synthesized for use in defense against attack or as attractants for pollinators.

We take advantage of the munificence of plants as chemical factories in many different ways, most crucially in medicine. Worldwide, at least a quarter of medicines come directly from plants; in some places, much more.

SIGNALING

Plants and animals put a great deal of time and energy into attracting attention to themselves. By way of colors, scents, sounds, vibrations, and elaborate movements, such as those in mating displays, living things send signals to one another. The distances over which these signals are sent may be short (a few centimeters) or long (kilometers in the case of some chemical sex attractants; the male gypsy moth can detect the scent put out by a female over 3 km away!).

Signaling is one of the essentials of life. No plant or animal could afford the cost of giving off "pointless" signals; the energy cost is too great. In *The Origin of Species*, Charles Darwin put it as well as anyone could in the case of flower color:

> Flowers rank amongst the most beautiful productions of nature; but they have been rendered conspicuous in contrast with the green leaves, and in consequence at the same time beautiful, so that they may be easily observed by insects. I have come to this conclusion from finding it an invariable rule that when a flower is fertilized by the wind it never has a gaily-coloured corolla [*petals*]. ... if insects had not been developed on the face of the earth, our plants would not have been decked with beautiful flowers, but would have produced such poor flowers as we see in our fir, oak, nut and ash trees, on grasses, spinach, docks and nettles, which are fertilized through the agency of the wind.

These conclusions are close to the opinions we have today: flowers are colored to attract mainly insects, prominent pollinators of flowering plants, and other pollinators, notably birds and some larger animal vectors.

What Darwin said for flower color is just as valid for other forms of signaling. Like any other activity, however, signals are fraught with dangerous side effects for both the sender and recipient. While signaling a prospective mate, an individual may attract the attention of a predator. The berries of some plants are colored red or black to attract birds, the outcome being the dispersal of seed. Other berries may be red or black to warn various animals that they are inedible, even poisonous. These signals may be confused by animals, with unpleasant or fatal consequences.

COLOR IN PLANTS

LEAF COLOR

A good place to begin a study of color in plants is in autumn leaves rather than flowers and fruits. The number of different colors in leaves is less than in the latter, yet the substances involved are the same in all cases; the gold of birch, the scarlet of sugar maple, and the crimson of oak have the same origin as the flowers of summer.

Yellow and orange colors: carotenoids

The simplest of the autumn colors are the yellows and oranges of birch, elm, poplar, and many other trees as well as most garden annuals and perennials. Pigments called carotenoids, as seen in the orange-colored root of carrot and the mature pumpkin, are already present in leaves in summer but are masked by the much larger amount of green chlorophyll. In autumn, however, deciduous plants begin their annual retrieval of useful materials from their leaves before they are discarded; one result of this scavenging is the breakdown of chlorophyll, which disappears, allowing the yellow and orange pigments to show through.

Carotenoids are found in all parts of plants, the annual production of which has been loosely estimated to be about 100 million tonnes, worldwide. Both animals and plants use small quantities of carotenoids as their source of vitamin A but, for reasons not always understood, many organisms seem to store more of them than can be justified for this purpose. In some animals, quantities of carotenoids taken in with food pass through the wall of the gut into fat and other body tissues; the golden color of cream or butter is one sign of surplus carotenoids eaten and stored by dairy cattle. Carotenoids that birds gain from eating berries and seeds are used to make beaks, legs, or feathers red, yellow, or orange, aspects of their signaling strategies.

Red and blue colors: anthocyanins

Red is the most spectacular color of autumn. Carotenoids can be red but most of the familiar red-colored autumn leaves and flowers have another pigment found less often in leaves during summer. Striking amounts of these pigments may be produced in dying autumn leaves to protect their activities until materials they contain have been moved out.

Cool bright weather in early fall favors the development of brilliantly colored foliage. The brighter the light, the more anthocyanin is produced by plants whose leaves normally turn red in autumn; the temperature must not be too high, either. In areas of the world where autumn is typified by mild, cloudy weather, autumn foliage is duller, limited to yellows and browns; the red anthocyanins are not a prominent feature of autumn foliage in these regions.

Thus, the range of autumn color is due mainly to only two families of pigments, carotenoids and anthocyanins, separately or together, in different mixtures, giving a range of subtly distinct hues.

F. S. Morot and the isolation of the first anthocyanin

The first flower pigment was isolated from blue cornflowers (batchelor-buttons) by the Frenchman, **F. S. Morot**, in 1849. The blue pigment Morot identified was named **anthocyanin** (a Greek word meaning *blue flower*), a word now used for a whole family of similar chemical compounds ranging in color from blue through purple to red and pink; these differences in color depend on other substances in petals.

FLOWER COLOR: WHY MORE VARIABLE THAN LEAVES?

In flower petals, carotenoids and anthocyanins combine to produce a wider range of coloration than is usual in autumn leaves. The crimsons, purples, blues, and creamy whites, as well as some reds and yellows, are usually members of the anthocyanin family; the oranges and other yellows and reds are carotenoids. The greater variety of color in flowers is caused by these pigments bonding with other compounds in petals, either simple minerals or more elaborate substances.

Example: blueness in flowers

One example of the link of color to the bonding of pigments is the wide range of blueness in flowers. Blueness in some flowers is caused by anthocyanin bonding to one of the tannins, which are not just found in places like tea (where we notice them as brown stains on cups or teapots) but are common to all plants. Other combinations can give different tones of blueness: anthocyanins linked to the metal, iron, produce the blue color of cornflowers; in *Hydrangea*, if a certain balance between aluminum and molybdenum in petals exists, flowers are blue; if there is a different balance, petals are red.

Pigment combinations in flowers

Pigments can combine with one another: the brown color of some petals in variegated wallflower is a combination of a magenta anthocyanin and a yellow carotenoid; variations in color in these petals are caused by accumulation of these pigments in different amounts in different places. The deep purple at the centre of some poppies is the result of higher concentrations of the pigments which give the delicate mauve at the extremities of the same petals. Concentration differences can be quite large; in purple cornflower varieties, anthocyanin concentrations are 30–50 times higher than in blue varieties.

White flowers: UV light reflectors

Finally, there are the pigments which produce the near white colors of many flowers. These are members of the anthocyanin family which seem to us to have no color except, sometimes, a creamy or ivory tinge. Although not spectacular to us, these pigments are clearly visible to eyes that can intercept UV light, such as those of bees and other insects.

COLOR PREFERENCES OF POLLINATORS

None of the pigments is there strictly for our pleasure; all have likely evolved for a reason. Evolution of color in flowers is far from haphazard, selection for it depending on the pollinators present.

Bees prefer blue and yellow flowers but can also discriminate among the types of anthocyanin pigments which to us appear white but which give off UV light. Bees are not especially sensitive to red colors but still visit some red-flowered species (poppies, for example), guided by pigments not visible to the human eye.

Other pollinators have their own color preferences. For instance, hummingbirds, sunbirds, and honeybirds are sensitive to red; butterflies are attracted to brightly colored blossoms; moths and wasps prefer duller, drab colors.

Bats, flies, and beetles are not attracted by color, depending mainly on other kinds of signals, such as aromas, to draw them to their host plants.

SIGNALS AT A DISTANCE

Quick recognition by migrating birds of the kinds of flowers that will provide them with food is critical. Since they pass rapidly from one area to another during migration they benefit from signals at a distance; it is not practical for them to spend much effort on time-consuming foraging for food.

Hummingbird preference

The color preference for fiery red among flowers pollinated by hummingbirds is an example of signals which function at a distance rather than at close quarters. Rapid recognition of possible food sources close by is especially critical to hummingbirds which, because of their large surface area and tiny volume, need a lot of food to compensate for energy lost as heat.

FLOWER COLOR AND HABITAT

In tropical habitats, where bird (particularly hummingbird) pollination is frequent, flowers are more often scarlet or orange. In temperate climates, more flowers are variations on blue, favored by bees and some other insects. This is not to imply that blue flowers are absent from the tropics or red flowers from temperate regions. Bees and other insects do, after all, operate in the tropics and birds act as pollinators in temperate regions.

Example: the phlox family

One good example of the trend to red in the tropics, blue in temperate climates, is in a family of plants whose members are found in both.

Members of the Polemoniaceae (the phlox family) are common in the Americas, some in the more northerly, temperate regions of the continent, others in central, tropical areas. Hummingbird-pollinated members of the family in tropical latitudes have scarlet petals; bee-pollinated flowers in the north are nearly all blue; others are mauve or pink and are pollinated by butterflies. In the 18 species in this family, all flowers pollinated by hummingbirds are orange–red, with one exception; all bee and bee–butterfly-pollinated plants are purple; butterfly-pollinated flowers are intermediate in hue, with various mixtures of orange–red and purple pigments.

Other color variations

Other examples demonstrate that at least some species can have flowers of one color in one place, and another color in an adjacent but different environment. In northern California some kinds of plants growing in open grassland pollinated by bees have yellow flowers; close by, in darker redwood forests, the same species are pollinated by moths, which pollinate at dusk or at night, and have less showy white or pale pink flowers.

These color differences are not as difficult to evolve as might be thought since the majority of flower colorations involve just the two groups of pigments, anthocyanins and carotenoids, already described above. By altering the relative amounts of members of the two pigment families, different mixtures and, thereby, different colorations can be achieved. Plants have considerable genetic flexibility to stop, start, increase or decrease the formation of these two families of pigments depending on the pollinators available.

MANIPULATING POLLINATORS

In addition to producing colors as simple signals, some plants have gone one step further by evolving ways to manipulate the behavior and movement of their favorite pollinators. Flowers in many plant families undergo color changes as they age, signals that can be interpreted by pollinators to the benefit both of themselves and their plant targets.

Fritz Muller

More than a century ago, Charles Darwin arranged for the publication of a letter sent to him by the naturalist, **Fritz Muller**, in which Muller remarked on a multicolored *Lantana* growing in the Brazilian forest:

> We have here a *Lantana* the flowers of which last three days, being yellow on the first, orange on the second, purple on the third. This plant is visited by various butterflies. As far as I have seen the purple flowers are never touched. Some species inserted their proboscis both into yellow and orange flowers; others … exclusively into the yellow flowers of the first day. This is, I think, an interesting case. If the flowers fell off at the end of the first day the inflorescence would be much less conspicuous; if they did not change their colour much time would be lost by the butterflies inserting their proboscis in already fertilized flowers.

As Muller observed, floral color change benefits both plant and pollinator. The color phase of the flower provides an accurate indication of its sexual status and whether it still offers any nectar reward to a potential pollinator. Plants that change the color of their blossoms also tend to have abundant nectar with which to reward pollinators when their flowers are young. As the flowers age not only do they change color but they also produce less nectar and pollinators learn to link change in flower color to nectar availability, thus saving the energy they would otherwise use in fruitlessly exploring flowers where there is no nectar reward.

Responses to color are learned, not inherited

One must be careful to say that the ability of particular pollinators to discriminate between flowers on the basis of color is not linked to the fact that they are the natural pollinators of these plants. This is not a case of co-evolution of plant and pollinator. New pollinators, which have not seen a certain type of plant before, when introduced to it, quickly learn which flowers offer reward in the form of nectar and which do not, using color as a cue. This is a learned response and not one inherited by the pollinator.

We must also be careful not to conclude too emphatically that a particular color is the signal for a certain status. Thus, red color in flowers does not invariably mean that there is a reward for a pollinator or even that the pollinator has to be a bird because the flower is red. For example, one type of South American *Fuchsia* flower is green with purple streaks when it is young and contains abundant nectar. Later, when the flower is older, the color changes to red but these flowers have no nectar. The red-phase flowers in this case are ignored both by natural (bellbird) and introduced (silver eye and bumble bee) pollinators. Current speculation about the redness of flowers and pollination by birds is that perhaps migratory birds use red, so readily seen by birds, as a long distance signal. Such a strong signal may be less advantageous where local birds are the pollinators. Birds have no programmed preference for red flowers; rather, preference is governed by the connection the bird learns to make between color and nectar reward. The precise color is of secondary importance.

HONEY GUIDES

Visible guides

Once a pollinator is at close quarters, other signals (again color but also often scents) take over and guide the visitor to the site of the reward. The so-called **honey guides** are often part of the color patterning of flowers, their object being to guide the pollinator to the centre of the flower, where sex organs and nectar are located. They are often found in bee flowers and may or may not be visible to the human eye. A yellow spot on the lip of an otherwise blue flower or a series of dots on the petals leading the pollinator down to the area where nectar, pollen, and ovaries are located act just like markings on a roadway.

Invisible guides

In other instances honey guides are invisible to the human eye but can be seen in the ultraviolet by insects. In black-eyed Susan (*Rudbeckia*), in daylight, the petals are uniformly yellow to our eyes. If viewed in UV light, however, outer parts of petals are bright, those towards the centre, dark. Members of the carotenoid family of pigments are responsible for the brightness of the outer parts of the petals in UV light and for the yellow color in daylight. The dark color nearer the center of flowers in UV light is due to anthocyanins.

Thus yellow carotenoid pigments, highly visible in daylight, are used to attract pollinators at a distance; anthocyanins at the center of the flower act as guides once the pollinator has been lured in close. Combinations of carotenoid and anthocyanin family pigments are common in yellow flowers. Of course, other types of honey guide also exist.

Darwin was partially correct

Broadly speaking, Darwin was correct: flowers are colored to attract pollinators, not just insects, as he supposed, but birds as well. Bats, flies, and beetles are also significant pollinators but rely more on the perfumes produced by flowers than their colors. Since Darwin's time, we have learned more about the roles colors play in providing signals to pollinators, not just at a distance but also at close quarters, as well as the pigments used to create colors (see Box 11).

BOX 11. IRIDESCENCE IN PLANTS

Biological iridescence, the change in hue of a surface when viewed from different angles, results from precise structural properties, not from pigmentation. It is used by many birds, fish, insects, and reptiles as a sexual or species recognition signal. It was the array of iridescent colors in peacock feathers which sparked the curiosity of seventeenth-century scientists about the nature of light, leading, much later, to an understanding of its wave characteristics. The phenomenon can be caused either by multilayering of materials (leading to reflection from or transmission of light through many surfaces), light scattering by crystals, or diffraction gratings at surfaces.

Iridescence is widespread in the plant kingdom and can give rise to vivid colorations. One example is the, aptly named, peacock fern (Selaginella willdenowii), *which is a startling, shimmering electric blue color. Blue iridescence is also found among marine (mainly red) algae and is widespread in angiosperms. As well as leaves, the blue color of some flowers and fruits has been shown to be caused by iridescence, not anthocyanin pigments. Despite the widespread occurrence of it in the plant kingdom, the mechanisms of iridescence and its possible value to plants are poorly understood. An exception is a recent, detailed investigation of floral iridescence in hibiscus and tulip petals.*

Hibiscus trionum *and many species of* Tulipa *have closely ordered striations in the waxy cuticles overlying their petal epidermal cells. These wavy surfaces are so precisely ordered as to form diffraction gratings with a grating periodicity of about 1.5 µm, resulting in the generation of intense light at the near UV and blue end of the spectrum. These wavelengths happen to be those to which the bee eye is most highly sensitive. To determine whether the iridescence in these test plant species had a causative link to bee pollination, its effect on the behavior of*

BOX 11. (cont.)

bumble bees was investigated. After the elimination of all other possible properties that might be luring the bee pollinators, it was shown convincingly that bumble bees can use the continuously changing hues of iridescence in these flowers to identify them correctly.

Although over 50% of angiosperm species have striated cuticles overlying their petals, it is not clear how many are patterned sufficiently regularly to produce diffraction gratings and, hence, iridescence. It is very likely that some such surfaces do so, however; others may influence pollination simply through their tactile attractiveness to insects. In birds and butterflies, it has been shown that structural color through surface patterning can enhance pigment coloration. Such interplay between structural patterning and pigments could also be used as a tool in a range of important signals by angiosperms.

FRUIT AND SEED DISPERSAL, AND COLOR

Another key role of color in plants is in the dispersal of seeds. As Darwin, again, observed in his book *Origin of Species*:

> that a ripe strawberry or cherry is pleasing to the eye as to the palate ... will be admitted by everyone. But this beauty serves merely as a guide to birds and beasts, in order that the fruit may be devoured and the matured seeds disseminated. I infer that this is the case from having as yet found no exception to the rule that seeds are always thus disseminated when embedded within a fruit of any kind if it be coloured of any brilliant tint, or rendered conspicuous by being white or black.

Importance of ripeness of fruits

It is desirable for a fruit not to be eaten until the seeds are ready for dispersal. The reds and purples of ripe fruit become so imprinted in "birds and beasts" that they will leave unripe fruit alone until the learned color signaling ripeness appears. Unripe fruits are often green, to camouflage them among leaves, bitter, to repel predators such as weevils and caterpillars, and tough. When seeds are ready for dispersal, fruits finally ripen to brilliant colors, again, the result of mixtures of carotenoids and anthocyanins.

As fruits ripen, they usually soften, and their sugar content increases, making the fruit more attractive to eat. Some produce subtle, "fruity," scents which attract bats, beetles, and monkeys. Birds either have a poor sense of smell or none at all, but fruit bats, especially, which often feed at dusk or at night, depend heavily on odor to guide them to food.

Birds, color, and seed dispersal

Fruit-eating birds are primary seed disseminators worldwide. Because seed dispersal is so important to plants, the criteria birds use in choosing fruits affects reproductive success of plants significantly. Birds have excellent visual acuity and color vision, and use color to locate and recognize ripe fruits just as they do flowers. A fruit's color conveys information about its quality, nutritional value, and digestibility, all of which influence a seed disperser's choice of the fruit.

However, the disperser must also have a means to interpret the information, in context. If we take our own interpretation of fruit as an example. We know from experience that a red raspberry is ripe to eat but a red blackberry is not. Redness, in itself, is not a sign of ripeness-to-eat. A green apple may be unpalatable, whereas a red or yellow one, ripe. A green pear may be fully ripe but a red or yellow cherry, still unripe. Nothing inherent in color signals edibility or ripeness just as in flowers it did not necessarily signal high nectar content. Color does allow birds to identify a fruit and receive information about its nutritional quality.

FRAGRANCES AND FLAVORS

It is often not always clear why plants produce the odors they do. For example, many have special scent glands on leaf surfaces which are full of volatile oils and which burst, releasing their contents into the air, at the slightest touch. Animals can be very sensitive to leaf odors but what advantage to a plant such lures have is often obscure (however, see Chapter 13). Some oils are known to repel browsers and predators by making a leaf less palatable.

FRAGRANCES AND FLAVORS ...

... pleasant ...

In the case of flowers there is not much doubt: a flower produces an odor (either fragrance, flavor, or both) to attract pollinators. The scents especially fragrant to human senses, like roses, lilacs, and gardenias, are most often those produced by flowers visited by bees and butterflies. Odor is especially important to night-flying pollinators such as bats and moths.

There are many other plant odors not linked to pollination. The sharp smell of cloves or cinnamon, the musky odor of thyme or sage, the fresh smell of a pine forest, the welcome odor of "freshly percolated" coffee, the distinctive flavors of lemons, oranges, and grapefruit, are all familiar to us but why plants produce them is sometimes still a mystery; they have not yet been linked to specific purposes.

... unpleasant ...

Not all odors and flavors are pleasant. Some, such as the skunk cabbage, resemble rotting meat. These, unpleasant to our senses, attract certain pollinators. Carrion and dung insects, for example, are attracted by smells and tastes resembling those released by carrion or feces.

... and complex

What is striking about all plant fragrances and flavors is their distinctiveness and complexity. In some cases, a single chemical may dominate a flower scent. More often, a fragrance comprises a mixture of many components. The smell of a rose is quite different from lilac, yet, in both cases, the odor is the result of many chemicals in

delicate balance with one another, not the product of just one substance. Some odors may be the result of a combination of dozens of
compounds in different concentrations which, together, result in a
certain, easily recognizable signature. Why a flower smells the way
it does is often difficult to analyze. One component of a certain scent
may reinforce a second and a third, producing a characteristic odor.

ESSENTIAL OILS
Chemists have attempted to analyze plant fragrances and flavors for
centuries. Most of the odoriferous plant compounds found in flowers are so-called **essential oils**. This is not a particularly good name
for them because most plant perfumes are not oils at all. The name
was given to them centuries ago and historical accidents are often
difficult to undo.

The essential oils of plants are complex organic chemical compounds that will not dissolve in water but which, like their *true* oil
and fat namesakes, are readily soluble in alcohol, ether, or chloro-
form. They are highly volatile essences: when released, they evaporate quickly into the air even at room temperature. The word *essence*
in regard to essential oils refers to the ease with which their smell
spreads through the air.

Terpenes: one prominent group of essential oils
Essential oils comprise close to 20 different chemical groupings. Some,
like those found in pine trees, spruce gum, and resins, are grouped
under the general name, **terpenes**. We have come across terpenes
before but did not identify them as such: the orange and yellow carotenoid colors of flowers, fruits, and leaves are also terpenes. Natural
rubber is, too, which illustrates how different from one another members of this group of chemicals can be (see also Chapter 13).

The various mixes of terpenes produced by evergreen trees are
so distinctive that one species can be differentiated from another by
smell. Others are less pleasant. The smell of wet sheep's wool is the
product of terpenes dissolved in the lanolin of wool, while the smell

of fresh cut liver is the result of a terpene released by damage caused by the knife.

SCENTS AND POLLINATION

Insects are sensitive to tiny concentrations of chemicals in the air. As already mentioned, the male gypsy moth can detect the sex scent put out by a female moth even at a distance of 3 km. Flower odors can be highly effective in attracting bees, butterflies, and other insects at concentrations which, to our senses, are not even detectable. Whether strongly or weakly scented to us, flowers synchronize the maximum output of their odors with the time when their pollen is mature and a flower ready for pollination. The output is even coordinated with the time of the day when the most favored pollinators are active. Thus, scent release is centered around midday for daytime pollinators and around dusk for pollinators active at night.

Unpleasant smells are an attempt to mimic the odor of decaying protein or feces to deceive carrion and dung insects, which normally feed on dead or waste organic material, into transferring their attention to a flower to aid pollination. These odors are often extremely offensive and have names to match: cadaverine (as in cadaver) and putrescine (as in putrid) are two well named examples. Others are fishy or rancid.

A "putrid" example: Jack-in-the-pulpit

A few plants that produce putrid smells have been investigated in detail and represent fascinating examples of this form of allurement. In the case of Jack-in-the-pulpit, the blade-shaped spathe (the "pulpit") that surrounds the central upright pillar, the spadix ("Jack"), opens at night. At the same time, the spadix warms up to a temperature close to 30°C, a burst of heat that speeds evaporation of the odor put out by the spadix and attracts dung beetles and flies.

When insects land on the surface of the spadix they slide into the flower where they become trapped because the sides of the spathe and spadix are glacially smooth. During the next day or so, in their frantic efforts to escape, the insects transfer pollen, causing fertilization. Once this is achieved there is an immediate alteration in the surface of the spathe, which becomes wrinkled. Now, trapped insects can find a foothold, climb out of the flower and escape, their service to the plant fulfilled.

DIVERSITY OF FRAGRANCES AND FLAVORS

'Mousy' scent
The carnivorous pitcher plant gives off a "mousy" scent attractive to flies and other insects. After enticing victims into the pitcher trap, the compound that produces the attractive odor also then paralyses them, allowing their digestion in a fluid at the base of the pitcher.

Hallucinogens
Some insects go so far as to "learn" to recognize the smells of individual types of plants. In so doing, they limit their attention to a small number or just one type of plant. There are even suggestions that a few plants produce hallucinogens or narcotics to which insects may become "hooked." A powerful attraction like this seems to operate in *Datura innoxia*, a relative of Jimson weed, where the nectar has hallucinogens dissolved in it. Observers have noted the drunken flight of the hawkmoth after visiting flowers of this species.

Sex attractants
Some flowers mimic the sex attractants of the insect they prefer. In one case, the sex attractant of the oriental fruit fly is given off in the scent of the golden shower tree, the release of which encourages the male of the species to visit the flower and pollinate it.

In another example, one of the *Ophrys* orchids produces a flower that resembles in shape and color the female of a certain type of solitary bee. The male bee is attracted by a scent put out by the orchid flower which closely mimics the sexual odor of the female bee. In attempting to mate with the flower, the male bee pollinates it. The attraction of the male to this particular flower is so powerful and so specific that the orchid has no need to produce any reward for the pollinator in the form of nectar; nor is the female bee attracted to the flower, only the confused, deceived male.

Orchids, the most sophisticated

Orchids have the most sophisticated scent regimes of all the flowering plants for attracting and encouraging pollinators. In one further example, certain bees living in the tropical forests of Central and South America use the scents produced by orchids as sex attractants rather than wasting resources to form their own. As part of their sexual ritual, the males of these species congregate into swarms; females are attracted to the swarms of males and mating occurs. While pollinating certain orchids the male bees collect plant odors on their bodies and use them to attract other males of the same species to form the swarms to which females are lured. Interestingly, different orchids attract different kinds of swarming bees; in other words, although all the types of these bees have the habit of swarming, not all kinds are attracted to all types of orchids.

SUMMARY

- Plant colors, fragrances, and flavors are distinctive, often complex and, sometimes, aimed at very specific animal targets
- In those associated with flowers and fruits, the principal aim is attraction of pollinators and seed dispersers
- However, plants produce especially fragrances and flavors for reasons other than pollination (e.g. in the case of carnivorous plants, to attract sources of food; in other instances, for defense purposes – see Chapter 13), some of which are as yet unknown; perhaps the exact targets for more of these substances will become obvious in time
- We often assume that plants play a largely passive role in their relationships with animals. But, if we can learn one thing here, it is that plants are able, through a variety of signals, to influence directly the behavior and movement of the animals on which they depend. They are not just a passive, pretty backdrop in a landscape occupied by animals but are consummate manipulators with dynamic strategies of their own.

BIBLIOGRAPHY: PART III

Amasino, R. (2004). Vernalization, competence, and the epigenetic memory of winter. *The Plant Cell*, **16**, 2553–9. *Historical review of the connection between exposure to cold temperatures and the promotion of flowering.*

Barinaga, M. (1998). Clock photoreceptors shared by plants and animals. *Science*, **282**, 1628–30. *Commentary on the fact that cryptochrome is found in animals as well as plants.*

Black, M., Bewley, J. D. and Halmer, P. (eds) (2006). *The Encyclopedia of Seeds: Science, Technology and Uses*. Wallingford, UK: CABI International. *Includes a review of the evolution of dormancy.*

Boysen-Jensen, P. (1910). Transmission of the phototropic stimulus in the coleoptile of the oat seedling. *Berichte d. Deutsche Botanisch Gesellschaft*, **28**, 118–20. In *Great Experiments in Biology*, M. L. Gabriel and S. Fogel (eds). Englewood Cliffs, NJ: Prentice-Hall, 1955. *Boysen-Jensen was the first to show that the phototropic stimulus in plants is likely to be of a chemical nature.*

Creelman, R. A. and Mullet M. E. (1997). Biosynthesis and action of jasmonates in plants. *Annual Review of Plant Physiology and Plant Molecular Biology*, **48**, 355–81. *A review of the jasmonates.*

Darwin, C. (1859). *The Origin of Species by Means of Natural Selection*. London: Murray. *Darwin concludes that flowers are colored when pollinated by insects but not when wind-pollinated.*

Darwin, C. and Darwin, F. (1955). Sensitiveness of plants to light: its transmitted effects. In *Great Experiments in Biology*, M. L. Gabriel and S. Fogel (eds). Englewood Cliffs, NJ: Prentice-Hall. *Excerpt from The Power of Movement in Plants, C. Darwin and F. Darwin. London: Murray, 1880, which highlights the pioneering work of the Darwins on the curvature of plants towards light.*

Davies, P. J. (2008). Bright days, cool nights help create autumnal splendor, says Cornell Plant Physiologist. *Science Daily*. http://www.sciencedaily.com (accessed July 11, 2008). *Discussion of what kinds of prior conditions lead to more or less brilliant autumn colors.*

Dodd, A. (2007). New research alters concept of how circadian clock functions. http://www.sciencedaily.com (accessed July 11, 2008). *A report that a molecule important for stress signaling in plants also regulates their circadian clock.*

Delph, L. F. and Lively, C. M. (1985). Pollinator visits to floral colour phases of *Fuchsia excorticata*. *New Zealand Journal of Zoology*, **12**, 599–603. *Links color change in flowers to nectar reward and pollination.*

Dukas, R. and Shmida, A. (1989). Correlation between the color, size and shape of Israeli crucifer flowers in relationships to pollinators. *Oikos*, **54**, 281–6. *Whether pollinators, in this case bees, differ in their innate color preference.*

Fankhauser, C. and Staiger, D. (2002). Photoreceptors in *Arabidopsis thaliana*: light perception, signal transduction and entrainment of the endogenous clock. *Planta*, **216**, 1–16. *Review of light-related developmental processes in plants, the photoreceptors involved, and the circadian clock.*

Goodwin, T. W. and Mercer, E. I. (1990). *Introduction to Plant Biochemistry*, 2nd edn. Oxford: Pergamon Press. *The role of flavonoids in plant coloration and how their color is influenced by factors like pH and metals.*

Harborne, J. B. (1993). *Introduction to Eological Biochemistry*, 4th edn. London: Academic Press. *Extensive discussion of the types and roles of color and scents in plant pollination.*

Jones, P. (2001). Smart proteins. *New Scientist*, 17 March, 2001, 1–2. *Jones outlines the structures of different kinds of proteins and the array of tasks they perform in the living cell.*

Kende, H. (1998). Plant biology and the Nobel prize. *Science*, **282**, 627. *Kende points out that ethylene in plants was the first gas shown to be a signaling molecule in biological systems, not nitric oxide in animals.*

Klein, R. M. (1987). *The Green World: An Introduction to Plants and People*, 2nd edn. New York: Harper and Row. *With chapters on the use of plants in human nutrition, medicines, and religion.*

Lamont, B. B. and Collins, B. G. (1988). Flower colour change in *Banksia ilicifolia*: a signal for pollinators. *Australian Journal of Ecology*, **13**, 129–35. *Bird pollinators learn to discriminate between flowers with nectar reward and those without on the basis of flower color.*

Lee, D. W. (2007). *Nature's Palette: The Science of Plant Color.* Chicago: University of Chicago Press. *Full coverage of color in plants, including iridescence.*

Lee, D. W. and Gould, K. S. (2002). Why leaves turn red: pigments called anthocyanins probably protect leaves from light damage by direct shielding and by scavenging free radicals. *American Scientist*, **90**, 524–32. *Authors hypothesize that plants protect their dying leaves so that retrieval of their contents can be completed before they are discarded.*

Leopold, A. C. (2000). Many modes of movement. *Science*, **288**, 2131–2. *Leopold outlines the fact that movements in plants are primarily based not on contractile proteins, as they are in animals, but on alterations to physical and cellular functions (e.g. change in shape or direction of growth).*

Lewington, A. (1990). *Plants for People.* New York: Oxford University Press. *Comprehensive account of the use of plants by humans, including how they cure our ailments and our use of them to adorn our bodies.*

Mas, P., Kim, W-Y., Somers, D. E. and Kay, S. A. (2003). Targeted degradation of TOC1 by ZTL modulates circadian function in *Arabidopsis thaliana*. *Nature*, **426**, 567–71. *Evidence for the control of circadian rhythmicity by gene loops forming an oscillator.*

Pain, S. (2002). Red alert. *New Scientist*, September, 41–4. *How plants use red anthocyanin pigments to protect themselves against damaging wavelengths of light.*

Pichersky, E. (2004). Plant scents: what we perceive as a fragrant perfume is actually a sophisticated tool used by plants to entice pollinators, discourage microbes and fend off predators. *American Scientist*, **92**, 514–22. *A comprehensive review of the use of scents by plants both to lure and repel.*

Sage, L. C. (1992). *Pigment of the Imagination: A History of Phytochrome Research.* San Diego: Academic Press. *Includes an account of the pioneering work of Tournois.*

Srivastava, L. M. (2002). *Plant Growth and Development: Hormones and Environment.* San Diego: Academic Press. *A comprehensive text covering many aspects of growth and development.*

Suarez-Lopez, P. and Coupland, G. (1998). Plants see blue light. *Science*, **279**, 1323–4. *Identification of blue-light receptors in plants.*

Trewavas, A. J. (1997). Plant cyclic AMP comes in from the cold. *Nature*, **390**, 657–8. *Definitive demonstration that, like other organisms and contrary to prevailing opinion, plants produce cAMP and use it to mediate the action of the growth substance, auxin.*

Troyer, J. R. (2001). In the beginning: the multiple discovery of the first hormone herbicide. *Weed Science*, **49**, 290–1. *Troyer cites Theophil Ciesieski's article: Untersuchungen über die Abwärtskrümmung der Wurzel. (Investigations of the downward curvature of roots.) Beitrage zur Biologie der Pflanzen, 1872.*

Tsukaya, H. (2003). Organ shape and size; a lesson from studies of leaf morphogenesis. *Current Opinion in Plant Biology*, **6**, 57–62. *Tsukaya discusses how leaf shape- and size-control mechanisms determine leaf shape at the whole organ level via cell–cell interaction.*

Weiss, M. R. (1991). Floral colour changes as cues for pollinators. *Nature*, **354**, 227–9. *A comprehensive review of how color changes influence the behavior and movement of pollinators.*

Went, F. W. (1955). On growth-accelerating substances in the coleoptile of *Avena sativa.* In *Great experiments in biology*, M. L. Gabriel and S. Fogel (eds). Englewood Cliffs, NJ: Prentice-Hall. *Excerpt from Went's article in Proceedings of the Koninklijke Nederlandse Akademie van Wetenschappen, 30, 10–19, 1926.*

Wheelwright, N. T. and Janson, C. H. (1985). Colors of fruit displays of bird-dispersed plants in two tropical forests. *The American Naturalist*, **126**, 777–99. *Color preferences of fruit-eating birds.*

Whitney, H. M., Kolle, M., Andrew, P. *et al.* (2009). Floral iridescence, produced by diffractive optics, acts as a cue for animal pollinators. *Science*, **323**, 130–3. *Demonstration, for the first time in a plant, of a causal link between iridescence in a flower and attraction of a pollinator, the bumble bee.*

Part IV   Stress, defense, and decline

Plants are exposed to unusual environmental conditions, daily and seasonally. Away from the equator, perennials such as trees and shrubs can be subjected to extreme cold in winter; plants growing at high altitude may experience, in addition to cold all year round (at least at night), drying winds and high levels of harmful ultraviolet radiation; desert plants must often suffer through long difficult periods of extreme high or low temperatures; in many locations, extended periods of drought or flooding may have to be endured; tolerance to increasingly saline soils may become necessary as we continue to abuse our arable lands; and soil, water, and air pollutants as a result of human activity may be encountered. **Stress** and how plants cope with it is the subject of the first chapter in Part IV.

Plants, both wild and cultivated, are surrounded by bacteria, fungi, nematodes, mites, insects, mammals, and other living hazards to their wellbeing, all hungry, many potentially harmful. Plants cannot easily avoid these enemies by moving away or hiding. The first part of the second chapter in Part IV focuses on the strategies plants use to combat the enemies in their environment, first and foremost, the ever-evolving **chemical warfare** that they wage against constantly adapting foes.

However, plants live in communities, as do other organisms, within which they compete with one another for moisture, light, and soil nutrients. Plants have evolved a variety of ways to create *Lebensraum* in their generally overcrowded world. The strategies adopted to provide some breathing space around themselves by limiting competition from neighbors are often only partially successful. The last portion of the second chapter in Part IV deals with the controversial topic of **allelopathy**.

The final chapter considers the issue of how **death** occurs, specifically how plants "get dead" and what we know about the processes, called **senescence**, leading to it?

12   Stressful tranquility

INTRODUCTION

Plants are exposed to unusual, even extreme, environmental conditions, daily, seasonally, or from time to time depending on where they live. Beneath the benign face of the natural green world, plants are waging battles constantly against difficulties posed by their environments. Because these stresses often lead to reduced health in plants, just as they do in animals, they are also of considerable interest to agricultural scientists. Stressed crops usually produce lower yields. Understanding how plants cope with and respond to environmental stresses (often called **abiotic** stresses to distinguish them from those caused by diseases and predators, which are **biotic** stresses) is, therefore, important to breeders whose job it is to develop crop varieties with resistance to stresses while maintaining high yields.

WHAT IS STRESS?

The word **stress** was used first by engineers to explain what happens when a force is applied to an object; **strain** is the change in the object caused by the stress. For example, an elastic band can be stressed by forcing it to expand; strain is how much the band is stretched by the force applied. Stresses and strains in the physical world can often be precisely applied and measured.

DEFINING "BIOLOGICAL" STRESS AND STRAIN

In a cultivated context
Anything that does not allow a plant to reach its full potential is a stress which will have a consequent strain, such as lower growth or seed production. Such a definition of stress and strain is useful in

agriculture where robust estimates of the full potential of a crop can be made over several seasons by planting the same crop in different climates and different soils. Once this full potential is known, an estimate of how much a crop falls short of that ideal can be calculated. But how can full potential be assessed for plants growing in the wild?

In an ecological context

Many plants found in what seem to be the most stressful conditions on Earth appear healthy and are often not found anywhere else. To take one of the most extreme examples from the world of microbes to make the point: a few bacteria flourish in hot springs at temperatures around 90°C but will not grow at 80°C or at much above 100°C. Are these organisms stressed at 90°C even though they grow better at this temperature than at any other?

There is, in fact, no perfect way of defining stress in plants, especially when they are growing in the wild. For the sake of argument it is possible to say that **stress is any outside influence that prevents a plant from functioning normally**. Of course, "normal" is, itself, highly subjective and varies from one species to another. Normal temperature in the case of bacteria living in hot springs is around 90°C; 80°C is stressful. Both of these temperatures would cause the instant death of just about all other living things, so both temperatures are stressful and yet normal at the same time.

This same dilemma applies to any set of environmental conditions. What constitutes stress for one organism may not be for another. For example, are the extreme environments found in deserts or on Arctic tundra stressful for plants that normally thrive there? "Normal" is a moving target when it comes to the environment.

SOME STRESSES PLANTS FACE

Whether or not we can define what stress is in plants remains a topic of debate. But some of the more obvious stresses are under intense investigation because of growing interest in the influence of climate change on plants (see also Part V) and the need for crops with increased resistance to stresses.

WATER STRESS
Water stress may be caused by either an excess or a deficit.

Flooding stress
Flooding stress, which usually affects roots, causes a lack of oxygen in the soil. This, in turn, lowers such functions as respiration and nutrient uptake in affected roots.

Water deficit, or drought, stress
Stress caused by a lack of water is far more common than that owing to water excess. The most common result of lack of water is desiccation damage within living cells. Cell protoplasts lose turgor and may shrink, osmotic changes which in themselves can lead to disruptions to how a cell operates. Membranes surrounding (plasma) and within (e.g. mitochondrial, chloroplast, tonoplast) cells may be damaged and become leaky, compromising the functions of the cell compartments they surround even if rehydration occurs.

LIFESTYLE RESPONSES TO DROUGHT STRESS
Injury caused by water deficit has such serious consequences for plants that they have evolved a wide array of lifestyle strategies to protect themselves against it.

Water deficit escapers
In some plants, understanding reaction to drought stress is easy – they avoid it. These **stress escapers**, which normally inhabit hot or cold deserts, germinate, grow, flower, and form dormant seeds in a matter of days (they are often referred to as **ephemerals** because of this lifestyle) following seasonal rains or during short periods of summer warmth. They have no special defenses against drought since they rarely confront any of the extremes of the climate in

which they live. They simply withdraw into dormancy when adverse conditions arise.

Drought stress avoiders

Alfalfa, palms, and mesquite, for example, survive desiccation as adult plants by sending down deep roots, thereby ensuring a secure water supply. Cacti, on the other hand, with their fleshy stems and leaves reduced to mere spines, avoid drought stress by taking in as much water as possible when it is available and storing it for the future. A cactus desert may look sparsely populated with lots of bare ground between each plant, but just below ground level is a jungle of shallow roots spreading in all directions through which moisture from even the slightest rainfall is immediately taken in and stored.

Plants do not have to be like cacti and other succulents to be efficient in conserving water. All manner of other strategies exist among plants to help reduce water loss and avoid the effects of desiccation.

Many kinds of desert plants have small leaves; some are thick and fleshy and used to store water as in cacti; others are flat with low volume but large surface area from which heat can be lost by convective air currents. Efficient heat loss helps lower leaf temperature and reduces the need to cool the plant by other means such as by evaporation, thus conserving water.

Other plants reduce water loss either by clothing leaves in protective hairs, which may also be silvery in shade to reflect light, or by shedding them during dry periods. The creosote plant is a good example of a desert species which sheds its leaves, shutting down growth during the months or years of drought it must sometimes endure.

LIFESTYLE RESPONSES TO TEMPERATURE STRESS

High temperature stress

In environments where there is both high solar radiation and high temperatures, plants have evolved leaf shapes, structure, and orientation which lower the influence of the Sun's energy. Reflective leaf hairs and waxes, leaf rolling, vertical leaf orientation, and production of small, highly dissected leaves all aid in minimizing heat absorption and maximizing heat loss.

Low temperature stress

The hardiest higher plants are Arctic and alpine species which encounter low temperature much of their lives, although they are often somewhat insulated from the elements by snow cover. Many survive through modifications to their growth form.

Some alpines remain small and low to the ground and are either covered in hairs or coated in wax. They are often light-colored, even to the extent of having grey–green leaves, to reflect sunlight containing damaging ultraviolet rays.

Alpine and Arctic grasses often grow in compact mounds, the bases of which form protective masses within which young shoots are protected. In the center of these masses the temperature may remain above freezing even when the air temperature round about is well below.

Not that "compact and small" is always the way plants protect themselves against the cold. High in the mountains of central Africa and the Andes of South America giant plants such as tree groundsels and lobelias grow upwards of 6 metres tall. The groundsels insulate themselves by having leaves covered in woolly hairs. Both grounsels and the lobelias also have leaves that fold up every evening around buds to protect them from low night temperatures. Some lobelias have another, odd, strategy for their protection: tightly packed leaves around buds form a bowl into which a watery fluid is excreted by the plant to a depth of several centimeters. At night, the fluid may freeze across but never to a depth of more than a centimeter or so. The main shoot tip is at the bottom of the bowl immersed in unfrozen fluid, protected from the cold.

SALINITY: A MOST DIFFICULT STRESS

Plants find some stresses extremely difficult to deal with, one of which is waterborne or soil salts.

Plants may encounter high concentrations of salts, especially **sodium chloride**, in coastal areas, like river estuaries and salt marshes, and in inland deserts. Some species have evolved to survive or even prosper in such conditions, but the number is small.

Increasing salinity of agricultural lands caused by heavy irrigation and overcropping is of great concern. Water delivered to the soil by irrigation contains a mixture of salts, which are left behind when the water evaporates; overcropping causes depletion of soil nutrients; both lead to increasing soil salinity. Once certain levels of salinity are reached, affected land must be withdrawn from agricultural production since most important crop species are very sensitive to saline soil conditions. In China alone, millions of hectares of arable land are classified as saline, mostly resulting from centuries of overuse.

Plant responses to salinity
Plants protect themselves against high levels of salts in three ways. Some, like the mangrove, which grows with its roots in saline water, take in only the amounts of salts they need, excluding the rest. Others take in excess salts but as quickly export them again into salt glands on their leaves. In these cases, the salts crystallize on the surfaces of leaves where they do no harm. Finally, some plants take in quite large quantities of salts and then "salt them away" in areas of their bodies, such as cell vacuoles, where they can do no harm.

Of the important agricultural crops, some, such as beans, soybeans, rice, and maize, have almost no tolerance of excess salts; tomatoes, cotton, sugar beet, and wheat will tolerate somewhat higher levels; barley is the most tolerant and will grow on land made saline by irrigation on which other crops will not.

STRESS FROM ENVIRONMENTAL POLLUTANTS
Much is heard these days about environmental pollution, some of which is natural in origin, like smoke from forest fires started by lightning strikes or gases and dust from volcanic eruptions. But most

are the result of human activity. Pollutants in air, waterways, and soil are almost too numerous to count; the damage they cause to ourselves, other animals, and plants is often undetermined. The two most serious groups of pollutants with which plants have to cope are **heavy metals**, found in soil and water, and **volatile organic compounds** in the atmosphere.

Heavy metals and remediation

All plants need certain minerals from the soil (see Chapter 4), but increasing numbers of soils also now contain other minerals, such as highly toxic heavy metals like **cadmium, lead**, and **arsenic**, put there by human activity. Mining and drilling wastes, paper mill effluents, and deposits from emissions of gases into the atmosphere from heavy industries and automobiles, are all leading to increasing levels and more widespread distribution of heavy metals in the environment.

As with other stresses, plant species differ in their ability to tolerate pollutants.

Some thrive on soils rich in **arsenic, selenium, nickel, chromium, gold, cyanide, cadmium**, and other contaminants. As in the case of tolerance to salts, a few plants are capable of excluding heavy metals altogether. Others take up the pollutants in large quantities and accumulate them to levels that would be lethal to non-tolerant plants. These so-called **accumulator species** are sometimes useful in soil remediation, that is, in the clean-up of contaminated soil. Periodically, the plants can be harvested, the pollutants along with them. Contaminants like **selenium, nickel, zinc**, and **lead** have been removed from polluted soils in this way.

Increasing numbers of industries also now practice more responsible remediation of lands they contaminate under legislation by governments. Accumulator species themselves are not harmed by the pollutants; they protect themselves by binding harmful minerals to specially designed binding molecules called *phytochelatins* produced by the plant only when they are needed to neutralize potentially harmful metals.

Volatile organic compounds

Major gaseous pollutants found in the atmosphere with an effect on plants include **carbon monoxide, sulphur dioxide**, various **nitrogen oxides, fluorides**, and **ozone**. Some of these are released during volcanic eruptions, but most are produced from human sources such as car exhaust emissions, metal smelters, and coal-fired power plants.

Most plants can handle and detoxify modest amounts of carbon, sulphur, and nitrogen gases and, to a limited extent, "cleanse" the air around them. Larger quantities of any one of these pollutants can cause major problems for plants. Photosynthesis is particularly vulnerable to upset.

Ozone

The ozone layer high in the atmosphere (the stratosphere) may be of crucial benefit to us for filtering out ultraviolet light from the Sun, but ozone at lower levels (the troposphere) is a great danger to plants.

Like other gases, ozone enters the plant through the stomata where it quickly breaks down to produce a number of highly toxic compounds, which are very aggressive and attack many natural, essential products in cells, rendering them useless to the plant. Photosynthesis is one of the most seriously affected processes. Lowering the efficiency of sugar production by the plant by damaging its photosynthesis in turn lowers the yield of agricultural crops, the growth of forest trees, not to mention the destruction of flowers, shrubs, and trees growing along roadsides and in parks in urban areas (see also Chapter 16).

STRESS TOLERANCE

Many of the modifications in plant structure and function in response to natural and imposed stresses mentioned above and others discussed in the following pages involve one or both of two types of tolerances to stresses that plants have evolved.

A plant species may be **adapted** to a stress, passing on the adaptations from one generation to the next through genetic inheritance. Changes to a plant produced by the adaptation may or may not involve changes to the structure of the plant which are either visible to the naked eye or under a microscope.

Alternatively, a plant may become **acclimated** during its lifetime. Gradual exposure to a stress allows a plant to adjust its internal functions so that it can continue to live and reproduce. Of course, the *capacity* to adjust in this way must be inherited. If there is no built-in capability to adjust to low temperatures, for example, then the plant will die if exposed to them. What form a response takes becomes obvious only if the plant is exposed to the kinds of stresses to which it and its ancestors have been acclimated.

Adaptation is the permanent, inherited changes that have occurred to plant form and function because of *long-term exposure* to stress. **Acclimation** is the changes a plant is capable of making to its form and function if stressed *during its lifetime* (Figure 12).

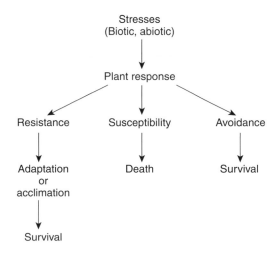

FIGURE 12 There are three ways in which plants may respond to environmental stresses: some remain susceptible and die; others may survive by avoiding a stress; yet others may have evolved mechanisms to either adapt or acclimate to the stresses imposed on them (after Hopkins and Hüner, 2009).

ADAPTATIONS TO WATER DEFICIT STRESS

Plants in many habitats likely face water deficit stress at some time in their lives. This has led to the evolution of a number of physiological strategies.

Opening and closing of stomata

The physical characteristics and role of stomata in the exchange of gases between plants and their environment was discussed earlier (Chapter 3 and Box 3). The critical role of stomata in maintaining the water status of a plant has been examined in great detail in studies of their physiological controls, which are numerous and elaborate. A summary of some important aspects of these controls can be found in Box 12.

BOX 12. CONTROL OF THE STOMA

Earlier (Box 3), the importance of stomata in regulating the exchange of gases (including water vapor) between a leaf and its environment was described. Control of opening and closing of the stoma is especially important during a period of drought stress and provides a line of defense against it.

*Movement of water into and out of guard cells regulates their turgor and determines the size of stomatal apertures. If loss of water by evaporation to the air is more rapid than its movement from adjacent epidermal cells into guard cells, turgor in the latter simply decreases, resulting in, so-called, **hydropassive stomatal closure**. When the entire leaf or the whole plant becomes dehydrated, **hydroactive stomatal closure** occurs. This results from the movement of solutes, then water, out of guard cells, leading to decreased turgor and closing of a stoma.*

The hydroactive process is triggered by even mild decreases in leaf water content and is linked to the low level of abscisic acid (ABA) synthesized continuously in leaf cells and stored in their chloroplasts. As water is lost from the leaf, some of this

BOX 12. (cont.)

ABA is released from mesophyll cells and transported to guard cells. As water loss continues, ABA is synthesized more rapidly in the entire leaf. In these two ways, the concentration of ABA in the guard cells rises, triggering partial or total stomatal closure.

The opening and closing of stomata is driven by the movement of K^+ and Cl^- into and out of guard cells together with the synthesis or breakdown inside these cells of organic anions like malate^{2-}, or the neutral sugar, sucrose. Changing levels of solutes such as these determine the direction and extent of the movement of water by osmosis across guard cell plasma membranes. A rise in solute levels inside guard cells attracts water, causing the cells to swell and stomata to open; a fall in solute levels leads to the opposite.

ABA has its effect on physiological responses in plants by binding to cell membranes. In the case of stomatal opening and closing, the binding of ever higher ABA concentrations to guard cell plasma membranes in response to drought triggers prolonged opening of anion channels in these membranes, permitting large quantities of Cl^- and malate^{2-} to flow out of the cells. This, in turn, causes opening of K^+ channels in plasma membranes, allowing massive quantities of this cation to escape as well. Together, these events promote stomatal closure. Increased binding of ABA to guard cell plasma membranes also inhibits the stimulation of stomatal opening by blue light; blue light enhances the movement of K^+ into guard cells and, hence, water influx by osmosis (see Chapter 3). These exquisite and complex controls underline the crucial roles of stomata in the physiology of higher plants.

Limits on photosynthesis

Of course, the rate of photosynthesis will be affected first by the extent to which stomatal openings are limited by water deficit. However, in addition, as water stress increases, dehydration of leaf mesophyll cells will also begin to lower photosynthetic capacity as cell metabolism is restricted.

Osmotic adjustments – osmolytes

As water availability decreases, cells throughout a plant may respond by raising their concentration of a variety of solutes, among which are certain metabolites specifically produced to maintain cell water potential without interfering with the normal functioning of cell metabolism. So-called **osmolytes** (e.g. **proline, sorbitol, mannitol, glycine betaine**) aid in maintaining cell water potential but do not interfere with normal enzyme function.

Changes in gene expression under osmotic stress

A group of genes activated under osmotic stress codes for proteins produced in embryos as seeds reach the stage of drying out. These proteins play a critical role in protecting cell membranes by binding water molecules to themselves very strongly, thereby increasing a cell's ability to maintain its water status.

Recent investigations using the most sophisticated molecular techniques show that up to 10% of all the genes examined change expression (higher or lower) in response to stress in plants. Some are under the control of abscisic acid (ABA), a plant growth substance known to be involved in responses to a wide array of stresses (see Chapter 10). In the case of osmotic stress, certain genes, all of them containing a particular nucleotide sequence, respond to the presence of ABA, triggering gene activity. Additional to these **ABA-dependent genes** is another set also influenced by osmotic stress, the **ABA-independent genes**.

The complexity of reaction to osmotic stress illustrated here is typical of responses to stresses in general in plants. Many physiological processes are involved in adaptations and acclimation to stresses and, hence, to the expression of genes regulating these processes.

ADAPTATION TO HIGH TEMPERATURE –
PHOTORESPIRATION REVISITED

The adaptations plants make to their structure in different climates are well known and are a major part of what is called plant ecology. Already discussed are ways in which plants are adapted to reduced moisture availability. Of special importance are tolerances some plants have evolved to withstand hot and arid conditions.

Photorespiration in temperate plants

As early as the 1920s, plants growing in temperate regions of the world were known to produce less carbohydrate by photosynthesis at higher than at lower temperatures. It was discovered that some of the carbon dioxide absorbed from the air was being lost again to the atmosphere at higher temperatures before it could be used in photosynthesis. When it was found that this process also involved consumption of oxygen it was given the name **photorespiration** (see also Chapters 1 and 2). Very high temperatures are infrequent in temperate zones of the world, where losses to carbohydrate production in leaves due to photorespiration are not large; not so in the tropics.

Special anatomy in some tropical grasses

In the 1960s, first in sugarcane, then in maize, it was discovered that certain grass-like plants which had their origins in the tropics differed from other grasses in their leaf structure. Rings or haloes of darker, denser cells could be seen surrounding the veins when thin sections through sugarcane and maize leaves were viewed under a microscope. Similar haloes were found in many other tropical and subtropical grasses but not in those from temperate climates. Later, a special form of photosynthesis associated with these grasses was linked to their different leaf structure.

C_4 photosynthesis

Tropical grasses like sugarcane and maize do not have photorespiration even at the high temperatures in which their wild relatives normally grow. These, and other similar grasses, have a more efficient way of delivering CO_2 from the air to where photosynthesis is occurring than do grasses found in temperate climates. Further, it was soon realized that this more efficient delivery of CO_2 was focused on the cells in the haloes surrounding leaf veins. It is there that carbohydrate formation in photosynthesis is concentrated in these species. CO_2 captured in all other parts of a leaf is directed to this small group of cells where sugars are formed with almost no loss of the gas back to the atmosphere.

The photosynthetic process in these tropical grasses has evolved to maximize the efficiency of:

- capture of CO_2 by the leaf;
- delivery of CO_2 to the cells in the haloes surrounding the leaf veins;
- movement of the carbohydrate formed in these cells directly into adjacent veins for transport to other parts of the plant.

Absence of photorespiration is an adaptation to the stresses found in the hot, dry places in which their ancestors evolved and survived.

Advantages of C_4 adaptation

Seasons of drought coupled with heat impose on plants the need to conserve water, which may force a plant to close, partially or completely, stomata in its leaves used to exchange gases with the atmosphere. But partially closing stomata also restricts intake of the CO_2 needed for photosynthesis. To compensate for this the leaves of some tropical grasses have evolved a mechanism to take into their photosynthesis apparatus every molecule of CO_2 available; they fritter away none of it in wasteful processes like photorespiration.

We know now that other kinds of plants also have this special kind of adaptation, or variations on it, to hot dry conditions, not just tropical grasses (see Chapter 1).

ACCLIMATION TO HIGH TEMPERATURE – HEAT SHOCK PROTEINS

Flowering plants cannot survive temperatures much beyond 50–55°C although many cacti and agaves tolerate 60°C routinely while a few have been known to survive short exposures to even higher temperatures. Even short exposure to higher temperatures induces most organisms to produce what have become known as **heat shock proteins** (HSPs).

HSPs are formed not just by plants but are ubiquitous also among animals and microorganisms. The amount of heat triggering their production does not have to be much above normal. Exposure, for a few minutes in some cases or a few hours in others, to temperatures just 5–15°C above normal is enough to trigger formation of HSPs. They appear with astonishing speed: the first hints of their formation can be detected within 3–5 minutes of exposure to heat; within 30 minutes they are produced in full amount; they disappear again within a few hours of relief from heat.

The HSPs formed by plants and most other organisms are numerous and can be grouped into five classes based on their molecular size. Different members of these classes are localized to different compartments of a cell (nucleus, mitochondrion, chloroplast, cytosol, etc.) and serve different roles in protecting the integrity of a cell's physiological processes.

ADAPTATION TO LOW TEMPERATURE
At the other extreme are plants adapted to life in freezing conditions.

Chilling injury
Susceptibility to chilling injury is common among tropical and subtropical plants but also occurs in many other plant species. When susceptible plants are cooled to temperatures below 10–15°C, their leaves become discolored, may even wilt, and show inhibition of photosynthesis, respiration, phloem translocation, and increased protein degradation. Especially common is the breakdown of membrane structure, leading to leakage of metabolites from leaf cells.

Freezing injury
Many plants, and animals too, can survive the more or less severe winter temperatures typical of Arctic, sub-Arctic, and temperate

regions. For example, insects living in these areas can manufacture glycerol when they are faced with conditions below 0°C, which acts as an antifreeze. In higher plants, adaptation to freezing conditions is more complex than this, but the outcome is the same.

Damage from cold is caused by ice forming between the cells of the plant. The problem with this is that water in solid form is useless to the plant, which suffers drought conditions as a result. Added to this may be the collapse of dehydrated cells under pressure from growing numbers of ice crystals in intercellular spaces. The water in the ice crystals originates inside cells, leaving them dehydrated and deformed, unable to function normally, the exact definition of a stress.

Supercooling

Frost-resistant plants have the capacity to defend themselves against low temperature stress in a number of ways.

Deep supercooling is common among those plants which inhabit climates where minimum temperatures in the winter reach −30°C or lower. The boreal forests across Canada and Russia, for example, are made up largely of conifers which display this kind of adaptation to low temperature stress. In deep supercooled tissues, water does not form ice crystals even at temperatures as low as −40°C. At temperatures much lower than this, ice formation cannot be prevented and damage to the plant will occur.

Although supercooling does not occur in all plants, it can frequently be found in the overwintering stems of large numbers of perennials in temperate zones and in many other plants that grow as far north as the tree line.

ACCLIMATION TO LOW TEMPERATURE

Boreal deciduous trees and shrubs like paper birch, trembling aspen, and willow, all of which can grow north of the Arctic circle, survive because they can acclimate to low temperatures.

During the growing season all suffer injury or death if exposed to even mild frost. Yet roots, shoots, buds, or other organs collected from these same plants after acclimation to low temperatures can be stored in liquid nitrogen at –196°C without injury. How this can be is being widely investigated, especially by agricultural scientists. Knowledge of how it is done could be of great significance to the development of frost-hardy crops. In some temperate regions of the world where the growing season is seriously limited by the number of frost-free days, knowledge of how to breed for frost hardiness is of great interest. At least two steps seem to be involved in acclimation in woody plants.

Frost-hardiness acclimation: stage one

In early autumn, in temperate regions when growth in perennial shrubs and trees is beginning to slow down but before leaves begin to fall, a signal is sent from the leaves to other parts of the plant. The signal, triggered by short days not by low temperature, is linked to the action of phytochrome, and is carried to all parts of a plant by ABA.

The role of ABA in cold tolerance is to trigger the synthesis of proteins, some of which are the same as those formed in response to water deficit and high temperature. The role of these proteins here is similar to that with other stresses: they aid in protecting the physiological processes of a cell.

Frost-hardiness acclimation: stage two

The second stage of acclimation is linked to the first frost, which is marked by many changes to the metabolism of the plant, including an accumulation of several sugars and proteins.

As temperatures fall further, increasing amounts of certain sugars accumulate in all parts of the plant. Most commonly, either **glucose, fructose, or sucrose**, or combinations of the three, build up in plant sap; less commonly, substances like **glycerol** and other antifreezes accumulate in some plants.

Lower temperatures also lead to the upregulation of genes responsible for the synthesis of those HSPs which protect against protein destabilization, thus aiding the maintenance of normal metabolism as the cell cools. In addition, cold stress triggers the upregulation of genes responsible for **antifreeze protein synthesis**. The particular role of these proteins is to inhibit formation and growth of ice crystals, thus limiting freezing damage.

In annuals, such as herbs and crop plants, acclimation to low temperature also involves ABA and production of particular proteins which appear only in tissues where cold stress is felt. Artificially adding ABA to these plants before they are exposed to low temperatures can protect them from freezing later and lead to the production of the same kinds of proteins as are formed naturally in response to low temperature stress.

OTHER STRESSES, OTHER PROTEINS

As more is learned about stress, scientists are discovering that similar, but not always identical, types of proteins are formed when plants are put under a variety of stresses, such as drought, oxygen deprivation, or saline soil conditions.

All living things have ways of protecting themselves against a wide range of environmental stresses. Some have developed very high levels of protection to survive in the harshest conditions. The cases discussed here provide only a glimpse of the range of proteins designed to protect against the environment should it become hostile, temporarily or permanently.

SUMMARY

- Stress in the plant kingdom is any outside influence which prevents a plant from functioning normally
- Plants face many stresses, some of the most important of which are:

  o water stress – either too much (waterlogging) or too little (drought);
  o temperature – either too high (heat stress) or too low (freezing stress);
  o soil or water salinity;
  o pollutants – on land, in water, or airborne;
  o volatile organic compounds;
  o ozone

- Plants may either adapt or acclimate to stresses:
  - adaptation is a genetically inherited capability arising from long-term exposure to an environmental stress;
  - acclimation is the ability a plant has to respond to a stress which may arise periodically during its lifetime
- Plants have evolved some remarkable and ingenious ways of responding to environmental stresses, both structurally and functionally.

So far, the resilience of plants in the face of these challenges, natural and imposed, has not been stretched beyond their ability to respond to and accommodate change. Increasing human activity is accelerating the severity of all the natural stresses faced by living things and is adding others not before encountered. Pollution on the scale we now see in some parts of the world was unknown even half a century ago. The consequences of that are all around us in our overpopulated cities and overcropped countryside. Hamlet's description of the air as "but a foul and pestilent congregation of vapours," is becoming true in some regions of the globe in a way that Shakespeare could not have imagined.

Given that the resources of living things to withstand environmental stresses surely have some limit, the increasing burden of human activity must now be regarded as one of the most serious challenges to plants. More about this topic follows in Part V.

13   Chemical warfare

Plants can be devasted by attacks from adversaries. Recall the near total damage to vegetation caused by plagues of locusts down the centuries or the wiping out of entire crops by disease as in the infamous example of the potato blight in Ireland in the mid-nineteenth century.

Yet, green plants still dominate the landscape in spite of their countless enemies; plants make up a major proportion of the world's biomass, the total weight of all living things. They have evolved an impressive array of strategies, physical and chemical, to defend themselves. Some plants even seem to use their weaponry to ward off competition from their own kind.

PLANT DEFENSES AGAINST PREDATORS

Indeterminate growth

Plants have an amazing ability to renew themselves even as they are being attacked. Grazing animals may spend major amounts of time cropping their preferred food sources, yet these same plants usually maintain healthy and vigorous growth as long as environmental conditions continue to be favorable. Disease may devastate a plant in the wild but rarely is the attack so complete as to wipe out an entire species. Renewal almost inevitably occurs, given enough time, because plants have an indeterminate growth style.

Physical defenses

Another partial answer to the question of why plants dominate the world is that many of them have developed effective physical defenses. Some are too tough and leathery for most animals; others have rapier-like **spines, thorns, prickles**, or unpleasant stinging or tasting **hairs** on their surfaces, making them altogether too formidable for most would-be predators to covet. Plants, in fact, seem to be generally unpalatable to other organisms. Only in selected cases does it seem that insects, grazing animals, and other organisms have the capability to overcome completely the wide array of defenses plants muster against them.

All plants exposed to the air are also coated with thin films of lipids (**cutin, suberin, waxes**), which not only reduce water loss substantially but also help limit invasion by pathogens (fungi, bacteria, viruses) and insect predators.

Chemical defenses

Prominent within the formidable arsenal of weapons is a bewildering number of chemicals that plants produce to help combat attack. Humans have understood for many centuries that plants can be hazardous to health. The lethal potion, hemlock (the alkaloid, **coniine**), given to the Greek philosopher, Socrates, was not the first time a plant product was used as a poison.

However, our view of plant toxins is rather narrow. We imagine that the only toxic plants are those that are dangerous to ourselves or our domesticated animals, but this is not so. Many others can be hazards to birds, fish, insects, and microbes even though to us they seem innocuous.

Plant products such as **nicotine, strychnine**, and **morphine** are certainly serious poisons to us, but we need to understand that most plants are more or less toxic even though they do not contain acutely dangerous poisons. Eat too much of any wild plant and it is likely to make you feel at least mildly ill. Attacks by animals and microbes are nothing new but have been going on throughout the billions of years that green organisms have existed. From their advent, plants have been faced with protecting themselves.

CHEMICAL CORNUCOPIA

Chemists in the nineteenth and early twentieth centuries began the formidable task of identifying and cataloguing a, seemingly, endless list of exotic chemicals (often referred to as **secondary metabolites**) which plants appeared to manufacture and other kinds of organisms did not. For the longest time, most of these substances were thought to have no function, to be merely waste byproducts. Some had been known for centuries for their use by humans as perfumes, flavors, drugs, poisons, and industrial materials (Box 13), but their value to the plants producing them was not so obvious.

BOX 13. VALUE-ADDED PLANT PRODUCTS

For thousands of years, the chemical compounds formed by plants mainly for their defense have been used by humans as medicines, for food preservation, and in religious rituals.

Some 2000 years ago, Dioscorides, physician to the Roman army, compiled a detailed account of several thousand plants in his De Materia Medica, *a book which remained the authoritative reference on medicinal plants for the next 15 centuries. Before development of modern food preservation techniques, civilizations depended on spices to ensure a supply of food, particularly during winter months. It was the search for a shorter (cheaper) route to the East Indies, source of many spices, which first led Columbus on his famous 1492 journey west from Europe to the Americas (hence the West Indies). Many references to spices, incenses, and other fragrant plant compounds can be found in religious texts such as the* Egyptian Book of the Dead, *the* Hindu Vedas, *and the* Judeo-Christian Bible.

It was the herbalists and apothecaries of the centuries from 500 to 300 years ago who were the forerunners of modern medicine. Over the years since, plants have given us heart drugs, analgesics, anesthetics, antibiotics, anti-cancer drugs, anti-parasite compounds, anti-inflammatories, oral contraceptives,

---

BOX 13. (cont.)

hormones, laxatives, diuretics, and many other aids. Even today, about one in four of all prescription drugs dispensed in our western societies have in them one or more ingredients derived from plants; in many other regions of the world the figure is more like four in five. The recent rise of interest in homeopathy may be an indication that herbal medicines have a future even in societies where they have been, for some time, viewed as relicts of a magical past.

Humans have used plant fragrances to enhance their attractiveness since earliest times. Legend has it that Antony was attracted to Cleopatra as much by her perfumes as by her beauty. Two of the three gifts brought to the infant Jesus by the Magi were frankincense and myrrh. The former is a gum produced by the olibanum *tree and was used by the Egyptians, Greeks, and Romans in incense. Myrrh is produced by a number of the* bdellium *shrubs native to Africa and south-west Asia and is used today in some cosmetics, perfumes, and pharmaceuticals.*

Development over 5000 years ago by the Egyptians of the art of perfumery led to the main use of plant fragrances by humans. The Egyptians used plant oils (olive, castor, sesame, jasmine) perfumed with herbs and gums to anoint their bodies. By the time of the Greeks and Romans, not only were perfumes used lavishly by men and women on their bodies, but clothes, bedding, and the walls of their houses were also anointed. Today, our use of perfumes and other fragrances continues to expand.

Among these compounds, we now appreciate, are many examples of chemicals plants use in their defense. Presumably, as plants evolved they progressively gained the ability to produce new kinds of chemicals. If some of these substances provided an advantage to the plant in some way (in warding off attack by predators, for example), then there would be a greater chance that the plant with the ability to manufacture a certain chemical would survive to maturity and pass on its new-found capability to the next generation.

Evolutionary defense and counter-defense

Not that predators then became helpless victims in the face of a new plant defense.

As new chemicals appeared in plants, insect pests, herbivorous animals, and disease microbes responded by evolving new, inherited strategies of their own to overcome the chemical obstacles faced. In turn, plants once more responded by evolving yet more diabolical chemical defenses, and so the battle continues. This is probably why some plants seem to have many chemicals which can be used to combat attacks by a variety of enemies while others have fewer. The speed and direction of the development of defenses was likely dictated by the challenges faced by each type of plant, in its native environment, down the millenia. The greater the challenges, the more protection accumulated.

Crop breeding lowers defenses

A consequence of their chemical defense strategy is that the compounds developed by plants to ward off attack also make them less desirable as food for humans. Wild ancestors are often not nearly as palatable as their cultivated descendants. Our crops have been bred to lower or eliminate unpleasant compounds helpful in the wild, making them more susceptible to insects and disease than their wild ancestors.

Toxicity is relative

Another point to remember is that how toxic a substance is depends on the dose size and how quickly it is taken in. Even water can kill if drunk in sufficient amounts over a short enough period of time (a condition called *water intoxication*). A dose of poison can be fatal if taken all at once but may not be lethal if imbibed a little at a time. The body may be able to handle something harmful if given enough time either to absorb it and deal with it effectively or begin passing it out again before the full, lethal, dose builds in the body.

An example of dose response to a poison is the case of the potato, which has in it a chemical, **solanine**, the amount of which is normally small, not a hazard. Potato tubers which grow above ground, however, and become green can have in them sufficient solanine to cause those who eat them to die of respiratory failure. The likelihood of death occurring depends on whether an individual has time to become accustomed to small amounts of the poison in the diet and can detoxify the substance as it is eaten.

THE DEFENSE COMPOUNDS
There are three groups of plant products within which examples of defense compounds are found.

Terpenes
Terpenes are a group which includes plant growth substances, such as **gibberellins** and **brassinosteroids**, as well as **rubber**, the **essential oils** (which lend the characteristic odors to foliage), and the main red, orange, and yellow pigments, the **carotenoids**, found in flower petals and elsewhere (see Chapter 11).

Many of the simplest terpenes are important in combating insect attack; for example, **pyrethrums** which occur in leaves and flowers of *Chrysanthemum*. Pyrethrums are popular insecticides because they are not harmful to humans or domesticated animals, do not persist in the environment for long, and are not serious contaminants of soils.

The **limonoids** are a group of terpenes well known as the typical bitter taste of citrus fruits. Perhaps the most powerful deterrent to insect feeding is the limonoid, **azadirachtin**, a product of the neem tree of Africa and Asia. In doses as low as 50 parts per billion, this potent plant terpene is a powerful insect control agent.

The **phytoecdysones** are steroids whose structure is the same as insect molting hormones. If an insect ingests one of these

compounds, progress through its normal molting instars is interrupted, often with fatal consequences.

In conifers, such as pine and fir, simple terpenes accumulate in
resin ducts in needles, twigs, and trunks, acting as repellents to such
pests as bark beetles. Others, in plants like sunflower and sagebrush,
are located in hairs on leaves and serve to repel anything with an
urge to devour the leaf.

Phenolics

Phenolics, the most familiar example of which is **salicylic acid**, from
which we produce aspirin, are used by plants in defense against
insects and fungi.

Some of the most intriguing types of phenolics are those
which act only when exposed to light, the **furanocoumarins**. When
activated by exposure to the UV-A (315–400 nm) region of sunlight,
these compounds bind to DNA, blocking its activity and repair.
Compounds of this type are found more often in foods like celery,
figs, parsley, and parsnips; if transferred to the skin during handling
in the presence of light they may cause rashes. In insects, the consequence of eating these compounds can be much more serious; the
insects may die if exposed to light after a meal. Some insects have
adapted to surviving on plants containing chemicals of this type by
living inside rolled-up leaves, where light is not bright enough to
trigger a reaction.

Next to cellulose, the commonest substances in plants are the
lignins, phenolic compounds which give wood its great strength.
Their physical toughness make them difficult for herbivores to
digest. Lignification also blocks the growth of pathogens and occurs
commonly at sites of infection or wounding.

Another group of phenolic compounds which resemble lignins
are the **tannins**, a word first used to describe compounds in plants
that would convert raw animal hides into leather through "tan-
ning." Tannins are very widespread toxins, which greatly reduce the
growth and survival of many plant-eating animals when added to

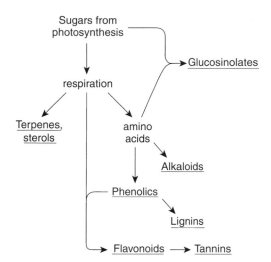

Sugars from photosynthesis

→ Glucosinolates

respiration

Terpenes, sterols

amino acids

Alkaloids

Phenolics

Lignins

Flavonoids ⟶ Tannins

FIGURE 13 The major groups of compounds that plants manufacture for use in chemical warfare.

their diets; they cause a sharp, astringent, sometimes unpleasant, sensation in the mouth. Unripe fruits, red wine, and well brewed tea all contain these compounds in abundance. In animals, they cause loss of appetite; in many microbes, they arrest growth.

Finally, **flavonoids** are one of the largest classes of plant phenolics, the most widespread group of which, the **anthocyanins**, are responsible for many of the red, purple, and blue colors in flowers and fruits (see Chapter 11). Other members of the group, the **flavones** and **flavonols**, absorb light at shorter wavelengths, are not visible to the human eye, but can be seen by insects such as bees, which use them as cues in seeking out flowers. They are also found in green leaves where they act to protect cells from UV-B radiation (280–315 nm) (see Figure 13).

Nitrogen-containing compounds

A major group active in plant defense are nitrogen-containing compounds, including the **alkaloids** like **nicotine, codeine, morphine, cocaine**, and **strychnine**, one or more of which are present in about one-third of flowering plants. All alkaloids, when taken in high enough dose, are toxic.

Other nitrogen-containing compounds also play a role in plant defense. Some are stored safely in the plant, but when an insect, slug, snail, or other plant-eating animal damages the plant tissues, they are released and produce **hydrogen cyanide** (HCN) gas which can cause the death of or, at the very least, illness in the attacker. Cyanide defense is common, being found in species as diverse as clover, sorghum, Lotus, bracken, and the desert plant, jojoba (see also below).

Mustard oils also produce volatile defense gases when released during attacks by predators. One subgroup, the **glucosinolates**, is responsible for the smell and taste of vegetables like cabbage, Brussels sprouts, broccoli, and radishes. They act as toxins or at least as very good repellents, as anyone knows who has tried to tempt young children to eat their vegetables! (Figure 13).

CHEMICAL DEFENSE STRATEGIES
Two main strategies are employed by plants in using toxic compounds to defend themselves.

STATIC DEFENSE
A single toxin, or a mixture, is present in tissues from the earliest stages of seedling development. In chicory, for example, three compounds are produced which are extremely bitter. Their effectiveness is increased by their inclusion in a sticky latex which oozes out of any damaged tissue. Other well known examples are deadly nightshade and hellebore, less well known ones like *Laburnum*, lupines, delphinium, and groundsel, and some surprises among our staple vegetables.

Cassava, a tropical root crop, must be specially treated to remove **cyanide** before it can become a food. Cyanide is also present in almonds, lima beans, and apple pips, sometimes in alarming quantities. In some vegetables, if, by law, the harmful natural compounds they contain had to be listed on packages, we would be forced to remove them from supermarket shelves in short order!

And then there is the **oxalic acid** in spinach and leaves of rhubarb. Fortunately, the amount in spinach is moderate, but rhubarb leaves are dangerous.

All of these, and there are many more, are examples of static defenses; there all the time in high enough quantity to act as deterrents. But maintaining high levels of complex chemicals in all parts of the plant at all times is costly in energy and materials.

VARIABLE DEFENSE

Deterrent compounds in vulnerable tissues
To moderate their cost, some plants have evolved a strategy of restricting toxin production to tissues most vulnerable to attack. In the coffee plant, **caffeine** is formed in young, juicy leaves that are attractive to predators to a level of about 4% of the dry weight of the leaf. As a leaf ages it thickens and toughens, making it less attractive to browsers; the rate of caffeine production decreases to just a fraction of a percent. Later, when the coffee bean has formed, caffeine concentration increases again to about 2% of bean dry weight before declining as the bean hardens and ages.

A more dramatic example of chemical protection concentrated in vulnerable plant organs occurs in the paper birch, a native of Alaska. A member of the terpene family accumulates to a massive extent in young, succulent parts of stems and twigs, areas of the plant favored as food by the snowshoe hare. The concentration of the chemical deterrent is 25 times greater in tissues favored as food than in more mature parts of the tree.

Mixture of deterrents
Another form of protection is to have different chemicals in different parts of a plant at different times in a growing season. Some plants have one chemical defense in the first leaves produced, another in leaves formed later in the year. This leads to leaves lower on stems and branches produced early in a growing season having different

chemical deterrents from younger leaves produced later, higher up the plant.

Insect hormone and pheromone production

Other forms of protection are more subtle and sophisticated than just production of unpleasant or lethal agents. First among these strategists are plants which produce and accumulate massive amounts of insect **hormones** (e.g. see phytoecdysones, above). Numerous examples have now been recorded of plants, especially conifers and ferns, that contain compounds which, when eaten by insects, upset the normal steps in their reproduction, leading to sterility and death. In one instance, large amounts of an insect hormone were isolated from the leaves of a species of sedge, enough to cause sterility in a wide range of insect predators, including grasshoppers.

Some plants mimic other substances formed by insects. An aphid that normally preys on wild potato secretes an **alarm pheromone** when attacked by a predator. This gaseous substance alerts other aphids close by that they are in danger and should take evasive action. The same substance is found in the hairs on the surfaces of wild potato leaves, which, when attacked by aphids, release the pheromone, driving away other aphids in the vicinity. This is a near perfect defense since a predator could never adapt to and ignore a signal so critical to its own survival.

Time and energy control of predators

Leaves of long-lived plants are particularly vulnerable to predator attacks. In trees, juicy leaves are produced year after year, maybe for centuries, time enough for insect populations to adapt to and overcome defense strategies. Obviously, determined insect predation could devastate a tree over time, which happens occasionally. Why not more often?

Patterns of insects' grazing show that only some leaves on any tree are heavily attacked; other leaves on the same tree are not attacked to the same extent. Leaves most affected have much lower levels of defense compounds than those less frequently grazed. Levels of tannins have been found to vary from leaf to leaf even on the same branch, yet one leaf looks the same as any other. All leaves have to be tested to discover which are palatable and which waste the insects' energy, making them more vulnerable to their own predators. In moving around in search of leaves fit to eat, an insect becomes more visible to a watchful enemy, which indirectly aids the plant by eliminating individuals in the predator population.

This variable defense strategy also works across a plant population. In the tropics there are tree communities in which some members have low levels of protective tannins; others have much more. An insect predator again must spend time and energy searching for low tannin individuals.

Variable deterrence is widespread

There is a heavy cost in materials and energy in providing protection as well. Thus, the incidence of variable deterrence is widespread in the plant kingdom. In the example of tropical trees, above, high tannin individuals compensate for the energy and material they use in providing their own protection by forming fewer leaves than low tannin neighbors. It is estimated that if half the leaves on a plant or half the individuals in a community of plants produce protective chemicals, there is close to 100% protection of the community. Put another way, if an insect is faced with a population of plants, half of which are unpalatable, but does not know which half, then it is not worth the time and energy it takes to seek out the unprotected individuals. Better to move on and find less well protected food sources.

INDUCED DEFENSE

Elicitors

Plants recognize the presence of an intruder by sensing molecules formed by the attacker, for instance, in the saliva of chewing insects. These elicitors can be any chemical unique to the attacker; proteins, carbohydrates, and fats have all been found to warn a plant of the presence of an enemy. The tiniest amount of elicitor released by an intruder anywhere into the plant body is sufficient to trigger a response by the victim.

Jasmonic acid and defense responses

Levels of the hormone **jasmonic acid (JA)** rise dramatically in response to damage caused by insects, triggering the induction of transcription in an array of genes involved in manufacturing a host of plant defense compounds. In some cases, JA has even been shown to accelerate the *primary* metabolic pathways, which provide the building blocks for *secondary* defense chemicals. This plant growth substance, JA, is emerging as a master regulator of defense metabolism.

Alarm agents

Induced defense in plants was first recognized about a half century ago but is now known to be widespread. In trees, this type of defense strategy usually involves quick production of tannins. In herbs, the toxin is more likely to be an alkaloid or a terpene.

One case in point is the attack by the hornworm on wild tobacco leaves where production of the alkaloids **nicotine** and **nornicotine** increases over 200% in roots, reaching a maximum about 10 days after invasion begins; the alkaloids are moved to the leaves as an attack continues. Interestingly, if attacked by nicotine-tolerant herbivores, plants do not increase nicotine but release certain so-called **green leaf volatiles**, which attract insect predators of their attackers. Clearly, these plants are able to distinguish which herbivore is causing damage and change defense strategy accordingly.

A wide array of green leaf volatiles are used by many kinds of plants to defend themselves against attacks by predators. For example, caterpillars grazing on tomato leaves do not go unnoticed. Unwounded leaves receive an alarm signal from a wounded neighbor, causing them to begin preparing for attack themselves. Long before they are invaded, leaves can begin manufacturing defense compounds.

Certain green leaf volatiles given off by plants under attack serve as signals to other members of the same species close by to begin expressing their defense-related genes. For instance, studies with the Sitka willow suggest that attacks by insects not only cause the leaves on the same tree to change and become less palatable to the attacker, but leaves of willow trees nearby also change.

INSECT USE OF PLANT DEFENSE CHEMICALS

Some insects have been able to adapt to the chemical defenses of plants and in some cases use them to advantage. One example of the latter is a grasshopper which feeds solely on plants of the milkweed family.

Milkweed manufactures several complex compounds that severely disrupt heart function. The grasshopper defends itself against attacks from other predators by spraying a poison from a gland; the spray has been shown to contain the toxins produced by milkweed. If these grasshoppers are given a diet free from milkweed, the poison content of their defense spray falls dramatically.

The classic milkweed/Monarch/Blue Jay case

A well known example of this same plant-to-insect defense is that of the Monarch butterfly.

In the 1960s, in Florida, it was noticed that Monarchs reared from eggs on milkweed plants, when eaten by Blue Jays, made the birds violently ill; those not exposed to milkweed had no adverse effect on the predator. Analysis showed that the unpalatable butterflies contained significant amounts of the toxins produced by milkweed but were unaffected by them.

The classic milkweed/Monarch/Blue Jay case has been studied intensively since its discovery, investigations that have led to other discoveries about the influence that plant defense chemicals have on animals. One consequence is that other kinds of butterflies protect themselves against bird attack by mimicking the coloration of the Monarch. These mimics, such as the Viceroy butterfly, are not poisonous but because they are look-alikes, they are under the umbrella of protection of the Monarch while avoiding the need to store the milkweed toxic chemicals or develop their own immunity to them.

One more interesting point associated with Monarch butterflies arises from observations made at their overwintering ground in central Mexico where they are the victims of black-backed orioles and black-headed grosbeaks which together account for 60% of Monarch deaths. These two bird species feed differently on their prey. The orioles, which are sensitive to Monarch toxins, pick the butterflies apart, carefully selecting certain parts of the body where no toxin is found. Grosbeaks are much less sensitive to the toxins and feed wholesale on Monarch carcasses.

No defense strategy is perfect

The case of the Monarch butterfly is a good illustration of the fact that, no matter how good a defense is, none can provide complete protection, all the time. Indeed, there are examples of insects becoming *dependent* on a plant's protective chemical for survival.

The beetle, *Caryedes brasiliensis*, is a predator of seeds of the vine-like legume, *Dioclea megacarpa*, found in the forests of Costa Rica. About 13% of the seed dry matter is made up of L-**canavanine**, an amino acid, which is also a potent insecticide. The beetle, however, has evolved a way to use the L-canavanine as its main source of nitrogen as it progresses through its larval stages. The insect larva forms urea from L-canavanine, ammonia from urea, and, then, most of the amino acids it needs for protein synthesis from ammonia.

DEFENSE AGAINST PATHOGENS

So far, emphasis has been on how plants defend themselves against predators. But plants also have the capacity to use chemicals to defend against attack from disease microbes. One point to remember, though, is that it is wild plants that are usually more completely resistant to microbial attack; their cultivated cousins may have had resistance bred out of them. Breeding for resistance to disease often involves the reintroduction of genes from wild relatives into crop cultivars.

LINES OF DEFENSE AGAINST PATHOGENS
Plants have several levels of defense against microbes, some of which are the same as those discussed used to combat predators.

Physical defenses
Tough waxy layers or thickening of epidermis (with **lignins** and **callose**) at the point of attack act as physical barriers to entry of microbes into a plant but rarely prevent disease entirely. More effective is the impressive arsenal of chemical protectants that plants are capable of producing.

Antimicrobial chemicals
The first chemical line of defense in some plants are compounds that are usually present at all times and are often the same terpenes and phenolics which defend against invasion by plant-eating organisms. They may occur at the surfaces of a plant or be found in deeper tissues. In a few cases they are so effective as to prevent microbial infection entirely, but in most cases this line of defense is an imperfect one which moderates but does not prevent infection.

A simple example of this type of defense is substances which release **cyanide**; a number of pathogenic fungi cannot detoxify cyanide and are quickly killed by it. Brassica vegetables, like cabbage, broccoli, and Brussels sprouts, are examples where wild relatives defend themselves in this way but cultivated varieties have had the capability bred out of them for reasons of taste. Diseases produced by fungi, such as powdery mildew, are much more prevalent among cultivated than wild brassicas for this reason.

The pungent, **sulphurous** compounds produced by members of the genus *Allium* (onions, garlic, leeks) are one other case where the chemicals are harmless to the tissues where they are stored yet are anything but benign when released after damage has occurred, as anyone can attest who has wept over a sliced onion. These, too, have anti-fungal and anti-bacterial properties.

Again, these barriers to infection are only partial and by no means universal. The anthracnose fungus is a serious disease organism in cereals throughout the world. Cereals produce a number of tannins and phenolic compounds when invaded by this fungus. In turn, the fungal spores produce a mucilage which binds these compounds, a response which neutralizes the defenses of the plant and allows infection to continue. Worse still, in a few cases, the compounds in question attract rather then repel invaders. A soil-borne fungus of the *Sclerotium* genus, for instance, depends on the release of the sulphurous chemicals produced by onion roots to trigger germination and then infection of these same roots.

The hypersensitive response

The hypersensitive response to attack by a pathogen is characterized by the rapid death of cells around the infection site. Through the creation of a dead zone adjacent to the point of attack, the pathogen is deprived of nutrients. Accelerated death of cells in the dead zone is preceded by the accumulation there of the signaling molecule, **nitric oxide (NO)**, as well as highly reactive oxygen species like **hydrogen peroxide** (H_2O_2), **hydroxyl radicals** ($\bullet OH$), and **superoxide anion** ($O_2\bullet^-$). Together, these potent toxins bring about the destruction of cell contents.

Phytoalexins

A major step forward in the search for an understanding of how plants combat disease came in the 1940s when it was observed that some plants formed certain chemicals only at the time of invasion by a disease organism. Specific chemicals, it was proposed, came into action to "ward off" (**alexos**) disease organisms from the "plant" (**phytos**). Many plant families produce their own kinds of **phytoalexins**; hundreds of these substances have been identified and the list continues to grow. Most are unique to a particular infection. Phytoalexins have several characteristics which set them apart from other types of defense chemicals.

Phytoalexins are produced in a region close to the infection site; they are usually effective against a wide range of fungi and bacteria. Once production has begun it goes ahead rapidly. About an hour after an attacker is sensed by the plant there is a dramatic increase in phytoalexin production; within 8 hours there can already be enough phytoalexin present to arrest the disease.

Another striking feature of phytoalexins is their diversity; there is no simple relationship between chemical structure and toxicity; no prediction can be made ahead of time as to what kind of inhibitor a plant will produce in response to an infection. Also, several substances are likely to be produced at infection, not just one.

Phytoalexin production is the most sophisticated, effective form of defense against disease of which plants are capable.

Other defenses

Many plants respond to fungal or bacterial invasion by walling off an infected area so that the spread of the disease is physically as well as chemically blocked. Other plants counter-attack by producing enzymes to destroy the microbe. But these defenses are marshalled more slowly and tend to be back-up to the more rapid response represented by phytoalexins.

THE ULTIMATE BATTLE

Despite all these barriers disease still occurs; some virulent strains of microbes are capable of breaching all resistance barriers of a plant. When this happens, the reaction of the plant is quite different from its previous responses.

The metabolism of the host quickens as it attempts to combat the advancing microbe, something analogous to the fever produced in humans by an infection. The microbe responds by producing toxins designed to poison the host, weakening its resistance, and giving rise to the typical symptoms of the particular disease, whether it be wilting or yellowing leaves, stunted stems, or distorted roots. The host plant, in turn, attempts to neutralize the microbial toxins by degrading them or diverting them to where they do less harm.

This is the ultimate battle for control between the plant and the invading microbe, and the outcome is by no means a certainty on either side. The primary defenses of the plant have been overcome by this time; the plant is fighting for its life. It is then a matter of whether the plant can raise the pace of its metabolism sufficiently to keep ahead of the infection or whether the toxins from the disease organism will prevail and overwhelm the victim.

ALLELOPATHY

Just as many plant species are able to defend themselves against attack from predators or pathogens, some seem also to be capable of limiting competition from their own kind.

COMPOSITION OF PLANT COMMUNITIES

In any meadow, forest, tundra, prairie, or desert, a variety of plant species will be found growing side by side. It is not difficult to think of reasons why they might all be growing in the same place, especially the fact that they are likely to have similar requirements for water, nutrition, temperature, and light. All communities, plant or other, are influenced by the adaptation species have to particular growth conditions.

Plants have evolved ways to create *Lebensraum* in their crowded world although the strategies they have adopted to provide this breathing space by limiting competition from neighbors are often only partially successful. Some simply overgrow their competitors; others produce a tight rosette of leaves close to the ground under which other plants cannot survive; yet others produce a dense canopy of leaves, limiting light availability below; and then there is a chemical strategy, called **allelopathy**.

ALLELOPATHY: WHAT IS IT?

Allelopathy is the ability some plants have to produce and then export into their surroundings chemicals capable of either slowing the growth of or outright killing their nearest neighbors (allel = one another; pathos = suffering). The chemicals involved typically are simple members of either the **terpene** or **phenolic** families. Their origin is uncertain. How plants might evolve substances to ward off attack by predators or microbes is not difficult to imagine, but how they might have evolved an ability to secrete chemicals into their surroundings to eliminate or slow down neighbors is another matter.

One possible explanation is that allelopathic compounds were once internal defenses against predators, which, at some point, began leaking into the environment and, by accident rather than design, affected other plants. Limiting competition fortuitously helped the plant producing the chemical to survive, set seed, and pass on its new-found ability to its offspring.

Alexander Fleming – penicillin

That some organisms have evolved the ability to produce chemicals that are excreted to limit the growth of neighbors has been known for a long time. One of the most famous examples is the discovery by **Alexander Fleming** of the toxic effect of a fungal contaminant, *Penicillium*, on bacteria he was cultivating in his laboratory.

Penicillium secretes a substance into its surroundings which inhibits the growth of other microbes in its vicinity, but not itself. Fleming noticed that the growth of the bacteria was stopped whenever they came close to the fungus. His curiosity about what the fungus was producing led to the discovery of the first antibiotic, **penicillin**. Many chemicals with a similar property to penicillin have been found in many different kinds of microbes, although few have the same potent antibiotic properties.

Early scientific observations in plants – Augustin de Candolle

That higher plants also have a capacity to influence other organisms in their surroundings has been long suspected. The Swiss botanist, **Augustin de Candolle**, in the third decade of the nineteenth century, was one of the first to record occasions in which chemical interactions between plants occurred. He noted that thistles inhibited the growth of oat crops; that spurges inhibited the growth of flax.

De Candolle not only made field observations but also carried out his own tests. He discovered that bean plants died if their roots were dipped in water in which the roots of other bean plants had grown and suggested that bean roots must excrete toxic chemicals into their surroundings. Similar types of tests were carried out by others through the early years of the twentieth century in which the effects on seed germination of various kinds of plants demonstrated the toxic effects of substances secreted into the soil by roots.

THE ADVENT OF ALLELOPATHY

The word "allelopath" was defined in the late 1930s, as a result of the work done by de Candolle and others, to mean the chemical interactions between all types of plants; some today would also include interactions between plants and the microbes that attack them, as we shall see shortly.

Black walnut tree

Allelopathy in plants was recognized by gardeners and farmers long before it was scientifically investigated. It was common knowledge that some plants thrived when planted together while other combinations were less successful.

The black walnut tree was known since the time of **Pliny**, some 2000 years ago, to slow down or even arrest the growth of other species nearby. In the 1920s, it was shown that seedlings of tomato and alfalfa died when planted within about 25 m of a walnut tree. The soil zone over which the tree had influence seemed to coincide with the zone occupied by the roots of the walnut. The natural assumption was that some sort of root secretion was responsible for the toxic effect; this turned out not to be the case.

In the late 1950s it was discovered that the toxic effects were due to a substance leaching from the above-ground parts of the walnut tree; leaves, stems, and branches. The toxin was shown to have a benign form inside the plant and to become lethal to other species only when released into the soil. The area of toxicity around the tree corresponded, therefore, not with the extent of the root system but with the area underneath the leaf canopy. Leaf and twig fall and rain caused toxic chemicals produced in the canopy to be washed into the soil in lethal dose.

Bristle brush

Early studies uncovered another antisocial chemical produced by leaves of bristle brush, which grows in the hot deserts of southwestern USA. Unlike its neighboring shrubby perennials, which have a wide range of other species growing in their vicinity, bristle brush is surrounded by a zone of bare ground. The toxic action of bristle brush leaves is specific, being lethal to tomato but not sunflower, barley, or itself. The fallen leaves of bristle brush retain their toxicity for a year or more if they remain dry; the toxin, a relative of **benzene**, is leached out of leaf litter by rain.

Wormwood

A most careful early study of allelopathy was carried out during World War 2 by a German (**H. Bode**) and a Belgian (**G. L. Funke**), scientists working with wormwood (*Artemisia absinthium*). Zones in the soil about 1 m wide alongside rows of the plant in which other plants would not grow were shown to be caused by a single chemical compound, **absinthin**, found in hairs on wormwood leaves. Rain washes the compound off the leaves into the soil.

Self-inhibition: guayule

There are examples of plants that produce a compound which inhibits others of its own species. The rubber plant, guayule, grows, like bristle brush, in the south-western deserts of the USA and produces a substance in roots which causes self-inhibition but does not affect other species. Guayule trees on the edges of rubber plantations grow better than those in the middle; extra watering or application of mineral nutrients do not help.

The guayule self-inhibitor was isolated and found to be **cinnamic acid**, a member of the terpene family, a substance so inhibitory that it stops growth of the plant at a concentration in the soil of only a fraction of one percent; tomato seedlings are only affected if treated with 100 times this concentration.

Cases like this have raised questions as to why a plant would produce a substance which is toxic to its own kind and not to others. What possible advantage could there be to such a strategy?

Establishing Lebensraum

One possible answer is the way in which shrubs are distributed under the conditions found in hot deserts. In this environment, individuals of a given shrub are widely and uniformly spaced in a seeming attempt to share scant supplies of water and nutrients. Perhaps it is an advantage to the mature guayule plant to produce a powerful inhibitor to maintain a wide spacing between individuals of its own kind, thereby lessening competition for food and water.

One of the most striking features of chaparral deserts is the zonation of herbs found around thickets of shrubs. Often, surrounding each shrub, are zones of bare soil 1–2 m in radius; herbs grow only outside these zones. As in the case of guayule, members of the terpene family are usually responsible for maintaining these zones.

Many desert shrubs are surrounded by an invisible cloud of chemical vapors which penetrate the soil under leaf canopies, remaining there until destroyed by microorganisms active only after rain has moistened their surroundings. Otherwise, the terpenes (one of which is the familiar **camphor**) accumulate in the soil, slowing down or stopping germination of many kinds of seeds.

Doubts about allelopathy

These and other examples suggest that allelopathy between higher plants is real and that many different chemicals are involved. What is doubtful is whether the production of any of these chemicals is a deliberate strategy evolved by a plant to limit competition. The more likely answer is that the chance formation of a compound with toxic properties is carried over to succeeding generations because it provides an advantage to a plant that can form the compound over a neighbor that does not.

The compounds involved in allelopathy are often formed by leaves and stems, not roots, a fact that has led to much of the criticism of allelopathy. Why so many of the chemicals said to be involved in allelopathy are produced in the above-ground parts of a plant when it would seem more effective for them to be formed directly by roots, is a question for which, so far, there is no fully convincing answer. Leaf and twig fall are the most important ways to deliver allelopathic inhibitors to the soil; an alternative method prevalent in arid conditions is through the release from leaves of volatile vapors into the atmosphere and from there into the soil.

COMPETITION

It is, of course, competition which lies at the heart of the allelopathic interactions of plants. Competition is, after all, one of the most general and important aspects of the relationships between plants. We should view this kind of chemical warfare as just one more weapon in the wider, constant battle among organisms to establish and maintain themselves and their kind in a community. But the defenses against others of their own kind, if they exist at all, seem tepid when compared to the fearsome arsenal of weapons evolved to combat predators and pathogens.

STRUGGLE FOR EXISTENCE: A CLASSIC EXAMPLE

The older view that the many exotic chemicals produced by plants are waste products which build up in the tissues because there is no way to excrete them has now been replaced; for many, maybe all, a role will be discovered for these substances. Organisms tend not to put energy and materials into processes with no function. Over billions of years, plants, unicellular and, later, multicellular, have evolved ways to defend themselves against grazing animals and disease microbes; those well protected are more likely to leave more survivors than those poorly defended. This trend ensures the survival to succeeding generations of the ability to form defense products.

In response, enemies have evolved ways to overcome defenses erected by the organisms on which they depend for survival. This "chemical warfare" continues as plants and their adversaries carry on an endless evolutionary tussle to survive; a classic example of the struggle for existence, one of which we continue to take advantage.

SUMMARY

- Plants may develop both physical and chemical defenses against predators and pathogens
- Especially potent is the cornucopia of secondary plant metabolites (terpenes, phenolics, and nitrogenous compounds), which provide formidable, but not insurmountable, barriers to predation or infection
- These defenses may be static (there all the time at a constant level), variable (there all the time but not at the same level everywhere in organs or tissues), or induced (activated only when an attack has been launched, such as in the case of the hypersensitive response to pathogens)
- Some plants may be able to limit competition from their neighbors by releasing chemical inhibitors into their immediate environment (allelopathy)
- Both the physical and chemical defense strategies are evidence of the struggle for existence between species which began the moment life arose on earth and continues, undiminished
- Through the years, humans have taken advantage of, especially, the chemicals produced by plants for many purposes.

14 Senescence and death

So far we have dealt with how plant systems work to promote, sustain, and preserve life. But what do we know about the processes leading to decline and death in plants?

We tend to imagine death as a process that begins at birth and progresses to an end, sometime. "Lifespan" is the maximum length of time an organism *could* live if all the conditions of life were at their most favorable; the human lifespan, for example, is about 120 years, but most of us do not expect to be around that long. "Life expectancy" more closely describes the reality. The question is not "How long *could* I live?" but rather, "How long can I *expect* to live?" which is dependent on prevailing environmental, social, and cultural conditions. In some parts of the world, human life expectancy may be only 30 or 40 years, about the same as it was some 2500 years ago at the height of ancient Greek culture, whereas in others we know it to be over 80 years. Disease, starvation, predation, accident, and polluted environments are just some of the hazards faced by all living things which affect how long they survive.

LIFE HISTORY STRATEGIES

All species share one basic aim in life: the survival of at least some individuals to reproductive age is crucial to the passing on of genetic traits to descendants. What is important is the different strategies living things have evolved to achieve this fundamental aim. These we call **life history strategies**; in plants they can vary widely.

Cloning

Some organisms seem to be immortal as long as they have a food supply and a favorable environment. Bacteria simply divide by binary fission; both "offspring" continue until they, too, divide, ad nauseam. Some animals, like sponges, have the ability to fragment, releasing single cells, each of which is capable of producing a new individual. Some plants, such as a number of the grasses which spread by underground stems, are able to grow indefinitely. Ancient colonies of these species, such as buffalo grass in North America, may have propagated themselves in an unbroken line since the end of the last ice age 15000 years ago. Some soil fungal mycelia may have perpetuated themselves in the same way for hundreds of years.

Other examples of a similar kind are some of the antique garden plants, which have been continuously transplanted over centuries using small pieces of the plant body (grafts, runners, cuttings, rhizomes). Some apple varieties and grape vines in Europe and cultivated olives in the Middle East have been propagated in this way, sometimes for thousands of years. In modern times, propagation of plants from single cells using sterile tissue culture techniques is now routine in horticulture and forestry.

We call organisms perpetuated in this way directly from the parent, **clones**; all clones are biologically identical to the original parent. At least in the case of clones, death is not an inevitable consequence of life, and the terms "life expectancy" and "lifespan" lose much of their meaning, therefore.

Life history strategies of plants

Clones aside, plants seem to have the widest range of lifespans of any of the kingdoms of living things. Some of the oldest living individuals are plants. The best verified cases are the bristlecone pines in California, some examples of which are over 5000 years old. At the other extreme are some desert plants, the aptly named *ephemerals*, which germinate from seed only after a significant rain, then grow, flower, set seed again, and die in a matter of a week or two.

Here, though, we will concentrate on angiosperms, the flowering plants, to illustrate plant life strategies.

FLOWERING PLANT STRATEGIES

Angiosperms, both monocots and dicots, fall into two groups: poly-carpics and monocarpics.

Polycarpics

On the one hand are plants capable of reproducing many times during their lives, most of which alternate between periods where they put all their energy into vegetative growth, producing more roots, shoots, and leaves, and reproductive growth, producing flowers and seeds. These **polycarpic** (literally, bearing fruit many times) plants are typified by such examples as trees, shrubs, perennial grasses, and those which have bulbs or other storage organs, like tulips, irises, and gladioli.

Monocarpics

The remaining flowering plants are **monocarpic**; they bloom and set seed only once in a lifetime. All plants that live for 1 (annuals) or 2 (biennials) years are of this type. Biennials, typically, grow veg-etatively in their first year before flowering and setting seed in the second. A few perennials are also monocarpic, the most spectacular examples being those which can grow for 100 years or more before flowering once, then dying.

The most conspicuous examples of monocarpic life strate-gies are agricultural crops, where whole fields of plants germinate, mature, flower, set seed, and die at much the same time.

The value of annuals

Much of the information gained about life and death of the whole plant has been through the study of annuals, which have two advan-tages over other plants. First, they have the potential to multiply and spread very quickly since they tend to produce large numbers of seeds in a short time. Second, their life strategy allows them to reach the next seed generation in only a few weeks or months; they survive adverse climatic conditions in a dormant or quiescent state in the soil (see Chapter 10).

Constraints of the annual lifestyle

Weighed against these aids to survival among annuals are a few constraints. For example, it is difficult for them to grow sufficiently big in one season to compete with perennials; they tend to be "shaded out" if forced to compete with longer-lived plants. In the wild it is difficult for annuals even to establish themselves unless there is relatively bare ground in which to do so. Growth conditions of this kind are created by removal of other plants through cultivation, by the activities of burrowing animals, or through the felling or burning of trees in a forest to produce clearings. Annuals are often the first colonizers to reappear after a forest fire, but their dominance does not last once longer-lived plants begin their comeback.

'Bet-hedge' strategy – typical of annuals

Because of constraints such as these on annuals, it is vital to their continued wellbeing to have built into their life strategies a healthy degree of "bet-hedging," especially in regard to seed germination. It would be suicide, for instance, for an annual to have all its seed crop germinate at the same time or in an environment where conditions were likely to deteriorate to the point that all seedlings would die. Dormancy strategies have evolved in annuals to spread out germination of a single seed crop over several seasons (see Chapter 10).

'Big bang' strategy – typical of biennials and some perennials

If annuals depend on bet-hedging for survival then plants which spend one or more years in a leafy condition before flowering once and then dying can be called "big bang" strategists. At one end of this spectrum are the strict biennials, like carrots, beets, and onions among garden vegetables, which flower and die in their second year. At the other extreme are long-lived species like the Hawaiian silversword which grows for about 7 years before flowering and then dying. A *Puya* from the Bolivian Andes flowers only once after about 100 years, then dies; some of the *Agaves* (century plant), talipot palms, and one of the bamboos do likewise.

'Die-back' strategy – typical of many perennials

Perennials which produce vegetative and reproductive growth every year for a few or many seasons are a particularly diverse group with a wide variety of life strategies. They are typified by the fact that they occupy a certain space year after year but die back to ground level at the end of each growing season. The advantage of this is that shoots are much less vulnerable to drought, cold, heat, and predators. They generally have storage organs below ground which see them through periods of adversity. The main disadvantage, compared to the woody shrubs and trees, is that at the end of each season they surrender their position in the canopy of plants competing for light; at the beginning of each growth season they must re-establish themselves by producing again all above-ground parts.

Weighed against this apparent disadvantage, however, is the fact that these leafy perennials need not divert energy and other resources to production and maintenance of permanent woody, supporting structures like shrubs and trees do. Many can also "move around" through formation of rhizomes, stolons, or shoots that arise directly from roots. Grasses and sedges, for example, can spread over large areas to form intricate "clonal" stands; alternatively, they can "jump over" barriers such as uninhabitable patches of ground by simply growing across them to more favorable territory beyond.

Larger perennials and polycarpics

Larger woody shrubs and trees tend to delay investment of energy and resources in reproduction until sufficient growth has taken place to ensure their survival. Once established, plants of this kind devote some resources periodically to flower and seed production; individuals that do not succeed in competing in the high canopy usually die without producing offspring.

Often, growth of individual parts of a tree or shrub is only loosely tied to what is going on in the rest of the plant. The supply of water and nutrients may be important but even restriction of these essentials does not necessarily cause irreversible damage to a plant's tissues. A tree may be aging but its growing tissues are not dying.

Thus, many larger perennials, polycarpic trees and shrubs, show the same kind of aging and death patterns as animals in the wild where few individuals live long enough to die of natural causes. There is a steady culling of individuals through infant mortality, disease, natural disasters, and the activities of predators and browsers. In the case of long-term survivors, a twig can be removed, rooted, and a new plant raised which has the same degree of vigor as a fresh seedling. The "aging" of ancient trees, for example, is most likely linked to regressive changes like decay of trunks after decades of wear and tear or the inability of canopies of leaves any longer to provide food for the ever-increasing volume of non-green tissues of the plants.

SENESCENCE: THE PROGRAM OF DEATH

All plant species, no matter which life strategy they follow, have a built-in program leading to death of the entire organism or some part of it; the program is called **senescence**. The shortest-lived, exemplified by the annuals and biennials but also including a few long-lived monocarpics, tend to have their lives limited by strictly programmed senescence, typified by a grain crop where individuals die, en masse, at much the same time. Perennials tend to have much less precisely limited lives and to die through gradual attrition as they age.

Senescence in plants shows a wide range of patterns. Senescence of the **entire plant** after reproduction is the most extreme case. Somewhat less drastic is the death of all above-ground parts (**top senescence**) at the end of a growing season, a form characteristic of plants with storage organs. Less drastic again is senescence of an entire leaf array at the end of a growing season, leaving stems and branches bare for a period of time (**deciduous senescence**). **Modified deciduous senescence** is the periodic fall of leaves and small branches characteristic of evergreens. Needles of many evergreens last about 4 years, although in the case of the extremely long-lived bristlecone pine, 30 years is more common. Still less drastic is the sequential dieback of leaves along a stem as they age, a development often found in annual plants (**progressive senescence**).

All these patterns point to a common conclusion; that senescence is under a control regime.

CONTROLS OF SENESCENCE

Gardeners and horticulturalists know that it is possible to extend the life of an annual by constantly removing flowers and fruits; the period of blooming in garden plants can be extended by removing fading flowers before they are replaced by fruit. In this way, it is possible to extend the life of monocarpic plants, most of which live only a few months or years. Soybeans, for example, which are annuals, can be induced to grow as high as 8 m over a period of 15 months by removing flower buds as they appear. Soybeans can also be prevented from flowering altogether by keeping them under artificial light in long days and short nights (see Chapter 9); preventing flowering and seed production causes soybeans to continue leafy growth for years. Some other annuals can be kept alive for decades as long as their flowers are prevented from maturing.

Observations like these suggest that control of events leading to death depends primarily on what is going on inside a plant linked to flowering and seed production, although we also know that environmental factors, such as day length and temperature, have a role in some instances.

Controls by competition for resources ...

A striking feature of monocarpic plants is the sharp shift in the movement of resources (minerals and carbohydrates, for instance) away from leaves, shoots, and roots towards newly formed flowers and fruits. Growth of roots and stems and production of new leaves often decreases in these plants, stopping altogether soon after flowering begins. Thus, one early explanation had it that the signal for onset of senescence was the shift in nutrients towards flower and fruit formation.

... Cannot be the whole answer

Further investigation showed that diversion of nutrients to flowers and fruits cannot be the whole explanation of senescence, however. Most plants do not form large enough numbers of flowers to cause massive diversions of nutrients. In some plants, leaf senescence still occurs even if flower buds are continually removed. In others, like spinach, where male and female flowers are produced on separate plants (dioecious species), formation of male flowers is just as effective in causing senescence as is female flower production, even though the formation of male flowers requires much less nutrient consumption than female equivalents. In many trees, flowering occurs even before leaf buds burst, yet senescence still occurs at the end of the growth season.

Clues from leaf senescence

The many physiological, biochemical, and molecular studies of leaf senescence, in particular, carried out over more recent decades have demonstrated that, during the process, the cells of the leaf follow a common, highly regulated pattern of coordinated changes in structure, metabolism, and gene expression. The fact that even the most casual observer sees the fading of the green color of the leaf as an early sign of senescence is no coincidence. The earliest change in cell structure is, indeed, the breakdown of the chloroplast, which contains chlorophyll and about 70% of the protein in a leaf. The assimilation of carbon in photosynthesis is replaced by the breakdown of chlorophyll and other large molecules like proteins, lipids, and nucleic acids.

This is not to say that *all* organelles of a cell are destroyed right away. Nuclei, for example, remain intact and active until the latest stages of senescence. The processes of breakdown of macromolecules requires the synthesis of a wide array of **hydrolytic enzymes**, such as **proteases** (to break down proteins), **lipases** (lipids), and **nucleases** (nucleic acids like the RNAs), all of which require that a cell's genetic program remain active. Mitochondria, the source of energy for cell activity, must also remain active throughout senescence.

Genetic controls

It has come to be appreciated that senescence is controlled by a genetic program found in each living cell. The genes induced during senescence, the **senescence-associated genes (SAGs)**, include those which encode the hydrolytic enzymes mentioned above, those responsible for the production of ethylene, and those involved in the movement of breakdown products out of the leaf. Many other genes have their expression decreased during senescence, the so-called **senescence downregulated genes (SDGs)**, whose products are no longer needed by the dying cell (see Box 14).

BOX 14. PROGRAMMED CELL DEATH

Programmed cell death (PCD) is an efficient mechanism for the suicide of those individual cells or whole tissues that are no longer wanted. In animals, it is known as apoptosis (falling away or off), a gene-directed process including loss of contact between cells, shrinkage of cytoplasm, membrane blebbing, patterned DNA fragmentation, and disassembly of nuclei.

PCD also plays a crucial role in many plant developmental processes, including in angiosperms: the senescence of leaves; the death of petals after fertilization; the loss of contents in xylem tracheids and vessels as they mature; the death of root cap cells as the root pushes through the soil; the disappearance of either male or, alternatively, female reproductive parts during the formation of unisexual flowers; the bringing to maturity of only one of several megaspores during flower development; the degeneration of the suspensor during embryo development; and the death of cells in the aleurone layer of the seed, releasing nutrients for the growing embryo.

PCD in plants also occurs in response to biotic and abiotic stresses. In plant–pathogen interactions, PCD serves as a defense

BOX 14. (cont.)

mechanism. For example, in the localized hypersensitive response (see Chapter 13), necrotic lesions form around an infection site, in which entire blocks of cells die to give rise to a barrier between a pathogen and the living host tissue. There are also abiotic stress responses. A good example of this is the development of aerenchyma under low oxygen (e.g. flooding) conditions, in which cells of the root cortex are induced to die, giving rise to large air spaces, enabling greater diffusion of air from the upper parts of the plant.

However, although there are intriguing similarities between them, the programmed patterns of degeneration characteristic of animal apoptosis are not so evident in plant PCD. Hence, that animal apoptosis and plant PCD have a common ancestral origin has not yet been rigorously determined. Certain characteristics of plant cells make exact comparisons with the animal process difficult. For instance, the presence of a cell wall makes disposal of the entire cell corpse by phagocytosis impossible, unlike in animals. As well, plant cells have a central vacuole (animal cells do not), within which components of the machinery generated for cell destruction (e.g. enzymes) can be stored, thus delaying the onset of cell destruction, possibly for long periods. This capability makes it possible to have different speeds and patterns of deployment of the machinery in each case of plant PCD. The clearest instances so far observed of similarities between animal apoptosis and plant PCD have been found in the detection of pathogens by their hosts and in the subsequent molecular responses to infections.

Growth substance involvement

If redirection of nutrients came to be seen as an inadequate explanation of why plants enter senescence, the involvement of growth substances is certain. Decrease in supply of **cytokinins** from roots is partially responsible for the onset of senescence, in leaves at least. Conversely, artificially maintaining cytokinin supply to leaves delays leaf fall.

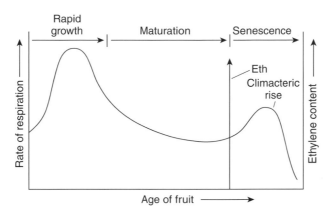

FIGURE 14 A rise in ethylene (Eth) concentration in an organ or tissue can act as a signal in senescence. For example, in some fruits, there is a spike in ethylene production at the point where the final, rapid phase of ripening and senescence begins. This phase is characterized by a sharp increase in respiration, the climacteric rise. In other cases, there may be no climacteric rise but ethylene may still be the trigger for processes leading to senescence (after Taiz and Zeiger, 2006).

Other growth substances, such as **ethylene** and **abscisic acid (ABA)**, accelerate senescence and death. In some fruits, ethylene production speeds up ripening and fruit drop (Figure 14; see also Chapter 7); in flowers, the commonest effect of ethylene is on the onset of wilting, followed by the fading of color, the export of nutrients, withering, and, finally, flower fall.

Taken together, studies of growth substance involvement suggest that senescence is regulated by the balance between ethylene and cytokinin with some further control, at least in the case of leaf senescence, by ABA. But the initial trigger setting the processes of senescence in motion is still unclear. Ethylene acts to increase the *rate* of senescence, not its *initiation*. For example, mutants lacking the ability to increase ethylene production show a delayed senescence, not an absence.

IMPORTANCE OF SENESCENCE

Whatever the trigger is for senescence, it is an advantage to having regulated programs for death and shedding of leaves, flowers, and fruits: shedding of fruits is advantageous in the spreading of seeds; in the case of old flowers, shedding leads to removal of a useless organ which could act as an entry point for infection; regular removal of old leaves allows renewal of a plant's photosynthetic apparatus.

Senescence can also be viewed as an aid to survival, not just a prelude to death. When senescence occurs, breakdown and redistribution of carbohydrates, lipids, proteins, and other large molecules mobilizes valuable nutrients which can be reused. Nutrient recycling helps certain plants, forest trees for example, to survive in soils where nutrients are in short supply. In this way, valuable nutrients can be recycled within the plant (see part V). Leaves shed in the autumn from deciduous trees in temperate climates would probably have little chance of withstanding cold winters so their loss, preceded by nutrient salvage, increases the chances of survival of individuals.

SUMMARY

- Plants have adopted a wide range of life history strategies for survival from one season or one generation to the next
- Prominent among these strategies are:
  - cloning (where the offspring are identical to the parent);
  - the monocarpic lifestyle (where a plant may flower only once in its lifetime, typical of annuals, biennials, and a few perennials);
  - the polycarpic life plan (flowering many times, found in most perennials)

- The kind of life strategy evolved by an organism depends very much on its environment. Whether it is better to breed once or many times depends on the chances of survival of offspring:
  - in environments where adult mortality rates are high (extreme climates, for example) or in which predators are numerous, breeding just once may be best;
  - less hostile environments may favor a multiple breeding strategy

- When a plant enters a path leading to death (either in whole or in part), it does so through a process called senescence
- Senescence is genetically and hormonally regulated.

BIBLIOGRAPHY: PART IV

Barz, W., Bless, W., Borger-Papendorf, G. *et al.* (1990). Phytoalexins as part of induced defence reactions in plants: their elicitation, function and metabolism. In *Bioactive*

Compounds from Plants, Ciba Foundation Symposium 154. Chichester: John Wiley and Sons. *Highlights the wide range of substances that can act as phytoalexins.*

Bennett, R. N. and Wallsgrove, R. M. (1994). Secondary metabolites in plant defence mechanisms. Tansley Review No. 72. *New Phytologist*, **127**, 617–33. *A comprehensive review of metabolites which play a role in plant defense.*

Collazo, C., Chacón, O. and Borrás, O. (2006). Programmed cell death in plants resembles apoptosis of animals. *Biotecnologia Aplicada*, **23**, 1–10. *Review of the resemblances and contrasts between plant programmed cell death and animal apoptosis.*

Crawley, M. J. (ed.) (1986). *Plant Ecology*. Oxford: Blackwell. *Strategies plants adopt to ensure their survival from one season and one generation to another.*

Engelberth, J., Albon, H. T., Schmetz, E. A. and Tumlinson, J. H. (2004). Airborne signals prime plants against insect herbivore attack. *Proceedings of the National Academy of Sciences USA*, **101**, 1781–5. See also Penn State (January 28, 2004). Signal chemical primes plants for pest attack. *Science Daily*, http://www.sciencedaily.com/releases. *Article deals with the release of green leafy volatiles (GLV) by plants used to turn on plant defenses in response to an attack.*

Gan, S. and Amasino, R. M. (1997). Making sense of senescence: molecular genetic regulation and manipulation of leaf senescence. *Plant Physiology*, **113**, 313–19. *An update on knowledge about senescence and underlying molecular controls.*

Harborne, J. B. (1993). *Introduction to Ecological Biochemistry*, 4th edn. London: Academic Press. *See below.*

Harborne, J. B. (1990). Role of secondary metabolites in chemical defence mechanisms in plants. In *Bioactive Compounds in Plants*, Ciba Foundation Symposium 154. Chichester: Wiley and Sons, 126–39. *In these two books, the author provides a wealth of information about the effect of the environment (air, water, soil) on plants, and plants on the environment, from a biochemical viewpoint.*

Leon, J., Rojo, E. and Sanchez-Serrano, J. J. (2001). Wound signalling in plants. *Journal of Experimental Botany*, **52**, 1–9. *Review of the mechanisms plants employ to generate wound signals.*

Leopold, A. C. (1980). Aging and senescence in plant development. In *Senescence in Plants*, K. V. Thimann (ed.). Boca Raton, FL: CRC Press. *Leopold addresses the issue of how senescence in plants is triggered, proceeds, and is regulated.*

Lesham, Y. Y., Halevy, A. H. and Frenkel, C. (1986). *Processes and Control of Plant Senescence*. Amsterdam: Elsevier. *Includes discussion of the plant senescence patterns proposed by Leopold.*

Macias, F. A., Galinda, J. C., Molinillo, J. M. G., Cutler, H. G. (eds) (2004). *Allelopathy: Chemistry and Mode of Action of Allelochemicals*. Boca Raton, FL: CRC Press. *Series of advanced articles on allelopathy, its efficacy and challenges.*

Newman, E. I. (1978). Allelopathy: adaptation or accident? In *Biochemical Aspects of Plant and Animal Co-evolution*, J. B. Harborne (ed.). London: Academic Press. *Addresses the controversy as to whether allelopathy is an adaptation or merely a fortuitous coincidence.*

Noodén, L. D. (1980). Senescence in the whole plant. In *Senescence in Plants*, K. V. Thimann (ed.). Boca Raton, FL: CRC Press. *That senescence involves deliberate, programmed changes which lead to death.*

Pennisi, E. (1997). Plants decode a universal signal. *Science*, **278**, 2054–5. *Demonstrates that the molecule cyclic ADP-ribose is a key element in a pathway that helps plants respond to stress and provides a commentary on other components of stress responses in plants.*

Pichersky, E. (2004). Plant scents: what we perceive as a fragrant perfume is actually a sophisticated tool used by plants to entice pollinators, discourage microbes and fend off predators. *American Scientist*, **92**, 514–22. *A comprehensive review of the use of scents by plants both to lure and repel.*

Ren, D., Liu, Y., Yang, K-Y. *et al.* (2008). A fungal-responsive MAPK cascade regulates phytoalexin biosynthesis in *Arabidopsis*. *Proceedings of the National Academy of Sciences USA*, **105**, 5638–43. *How a certain plant initiates a phytoalexin pathway in response to attack by a particular pathogen.*

Rosenthal, G. A. (1986). The chemical defenses of higher plants. *Scientific American*, January, **254**, 94–9. *Strategies and compounds used by plants to defend against attack.*

Skelton, P. (ed.) (1993). *Evolution: A Biological and Palaeontological Approach.* Wokingham, UK: Addison-Wesley in association with the Open University. *The variation in life expectancy among organisms, and reproductive strategy.*

Slater, A., Scott, N. and Fowler, M. (2003). *Plant Biotechnology: The Genetic Manipulation of Plants.* Oxford: Oxford University Press. *A useful overview of the field of genetic modification of plants, especially in regard to engineering stress tolerance.*

Wildon, D. C., Thain, J. F., Minchin, P. E. H. *et al.* (1992). Electrical signalling and systemic proteinase inhibitor induction in the wounded plant. *Nature*, **360**, 62–5. *Electrical alarm signaling in a wounded plant in response to attack.*

Part V   Plants and the environment

There is increasing alarm around the world about how human activities are forcing the pace of change in the environment. A major concern is the effect we are having on plants.

Plants are major regulators of the environment through their ability to capture energy from the Sun and convert it to a form available to other organisms; through their influence on shaping the composition of our atmosphere by removal of carbon dioxide and addition of oxygen during photosynthesis; by their retention, circulation, and evaporation of water below, at, and above the Earth's surface; and in their ability to mine and redistribute mineral elements in soil via their roots, thus contributing to the weathering and breakdown of the surface layers of the planet. All of these critical functions of plants are under threat because of stresses imposed on them through global warming, deforestation, afforestation, agriculture, industrial and domestic pollution, irrigation, drainage, and flooding.

Before the Industrial Revolution began a few hundred years ago, the pace of changes to the Earth was affected by human activities in only limited, localized ways. **Chapter 15** is devoted to an examination of this phase of development of our planet, particularly the cycling and recycling of some key chemical elements on which living organisms depend.

As industrialization took hold and grew, the pace of alterations to the Earth's environment began accelerating to what has now become an alarming rate, leading to changes to our land, water, and atmosphere resources, causing an increasingly wide array of stresses on plants. As we have seen in previous chapters, stress in plants manifests itself through a multitude of changes at the physiological level. Included in **Chapter 16** is an examination of some of these physiological effects and, hence, why it is important that we continue to increase our understanding of the fundamentals of how plants work.

15   Elemental cycles

INTRODUCTION

For nearly all of its four and a half billion years of existence, the Earth's environment has followed a path and pace of change affected much less by human activities than by other, much larger, natural forces. Humans began having an impact a few thousand years ago through such activities as the slash and burn gardens in the tropics and the burning of grasslands by hunter/gatherer societies prior to forest clearing for agriculture in temperate regions. In the past few hundred years, however, burgeoning human populations and their activities in all parts of the globe have greatly increased the scale of impact.

The goal of this chapter is to examine the period prior to the advent of widespread human effects on the Earth's environment, especially factors important to plants and the roles they play in the biosphere.

For the first half billion years or so after it began forming, the Earth was shaped and reshaped by physical and chemical forces alone. Around 3.8 billion years before the present (BP), the outer **crust**, which formed as the Earth cooled sufficiently to form a solid, unbroken surface layer, cracked into a number of massive plates floating on a deeper **mantle** of molten rock. Ever since, these **tectonic plates** on which the continents are carried have continued to move from place to place at the Earth's surface and, at their edges, to sink beneath one another into the hot mantle below, a process called **subduction**. In the fierce heat of the mantle these rocky materials are transformed to liquids and gases, which are then recycled back to the surface and atmosphere, under pressure, through the outpourings of volcanoes and uplift of mountain ranges. Earth is unique in the solar system in having tectonic plates, causing a continuous cycling and recycling of its vast stores of chemical elements, including those essential for life.

The **oceans** developed on the primitive Earth about 4 billion years BP, a milieu in which, a few hundred million years later, the first life likely appeared. There followed a several billion-year period of virtual stasis during which the living world was made up exclusively of microorganisms, first prokaryotes, then, also eukaryotes, unicellular and multicellular. About 400 million years BP, multicellular green plants evolved, the advent of which altered, drastically and uniquely within our solar system, the path our planet's environment took. Cycling of the Earth's physical resources through the atmosphere, on land and in the ocean, included, henceforth, increasing influences of plant life.

CYCLING MATTER

GEOCHEMICAL CYCLING

In the earliest period of Earth's development, **sediments** settling to the floor of the ocean were subducted into the molten mantle along with the rocky material of the tectonic plates (see above), returned to the surface through volcanoes and mountain uplift, dispersed at the surface by wind and water, returned to the ocean as dissolved and suspended matter in rivers or directly from the air to end up, again, as sediments on the ocean floor. This ancestral **geochemical cycling** continues today, a slow dance, the tempo of which is measured in millions or hundreds of millions of years.

BIOGEOCHEMICAL CYCLING

The appearance of life on Earth, possibly as far back as 3.8 billion years ago, led to a livelier tempo, **biogeochemical cycling**, a phenomenon quite possibly unique in our solar system. The concept of biogeochemical cycles was developed over 100 years ago by a young professor at Moscow University.

Vladimir Ivanovich Vernadsky and the
concept of the biosphere

In his book *Cycles of Life: Civilization and the Biosphere*, Vaclav Smil recounts the story of an English naturalist, G. T. Carruthers, who, in November, 1889, while onboard ship in the Red Sea, encountered an impressive swarm of locusts. In a flight of imagination, Carruthers estimated the total number of insects to be 24 quadrillion, weighing about 43 billion tons, a calculation he published in *Nature*, a report noted by **Vladimir Ivanovich Vernadsky**, who, over 30 years later, wrote in his book, *La Geochimie*, Smil notes:

> The swarm of locusts expressed in terms of chemical elements and in metric tonnes [*is*] analogical to a rock formation, or, more precisely, to a moving rock formation ...

As a mineralogist, Vernadsky imagined the rampage of the locusts as a way for chemical elements to be dispersed across the globe much more rapidly than by geochemical processes alone. He realized that plants were also involved since it was from them that the locusts gained their chemical elements. Further, he understood that plants, in turn, had taken in these chemicals as nutrients through their roots from the soil, part of the Earth's crust. Here was a biogeochemical cycle in action:

$$soil \rightarrow plant \rightarrow locust \rightarrow soil$$

Using this and other observations, Vernadsky went on to elaborate the concept of the **biosphere**, a term first coined in 1875 by an Austrian geologist, Eduard Suess (see Box 15).

BOX 15. VERNADSKY'S LEGACY

V. I. Vernadsky (1863–1945), although largely unknown in the west until recently, was one of an influential group of Russian scientists at that time. Included also were D. I. Mendeleev (author of the periodic table of elements), V. V. Dokuchaev (founder of

BOX 15. (cont.)

modern soil science), I. P. Pavlov (of "Pavlovian response" fame), S. Winogradsky (discoverer of chemotropism in bacteria) et al. Some consider that Vernadsky should be placed alongside Darwin in his significance to an understanding of the biosphere. The latter demonstrated the continuity and complexity of life through time; the former showed the unity of life in space and its significance as a geological agent.

Among other topics, Vernadsky studied the distribution and movement of chemical elements in the Earth's crust (lithosphere), in water (the hydrosphere), and in the atmosphere. These interests led him to investigate which chemical elements were found in living organisms and to the realization that almost all of them were present there. He went on to define the boundaries of the biosphere, above, at, and below the surface of the Earth. The dynamic nature of the biosphere and its continuity over billions of years led him, further, to elaborate on the concept of the biosphere as a force in cycling, recycling, and depositing chemical elements. For example, in considering the influence of marine organisms he emphasized their role in forming deposits of calcium carbonates, phosphates, and silicates. He also recognized the role of biochemical reactions in the concentrations of manganese and iron (BIFs) in the Earth's crust.

All this led Vernadsky to conclude "that the deposits of marine mud and organic debris are important in the history of sulfur, phosphorus, iron, copper, lead, silver, nickel, vanadium and, perhaps, other rarer metals." Further, he opined that the "gases of the entire atmosphere are in an equilibrium state of dynamic and perpetual exchange with living matter [which can therefore be viewed as an] appendage of the atmosphere."

Besides recognizing qualitative aspects, Vernadsky also attempted to quantify biospheric processes. His conclusions vary from recent data but his creation of global budgets for

BOX 15. (cont.)

biogeochemical cycles was highly innovative at the time. His idea of organisms as a major geological force led him to derive the expression "kinetic geochemical energy of living matter" to emphasize the enormous biogeochemical potential of the biosphere.

Latterly, Vernadsky recognized the ability of humans to transfer and concentrate chemical elements in the biosphere to an unprecedented degree. His last work was dedicated to the noosphere, his concept of the next stage of evolution of the biosphere driven by humanity as the dominating force.

THE IDEA OF THE BIOSPHERE

Carruthers' calculations of the size and mass of the locust swarm he observed were wildy inflated, but his observation was important to the development by Vernadsky of the role of organisms in the biosphere, not just in cycling and dispersing the Earth's chemical elements but also in maintaining them in a state of balance (homeostasis) in the environment, favorable to life.

The domain of life, the biosphere, has quite narrow limits, extending upwards only about 6–7 km above sea level, beyond which it is too cold and the air too thin for life to thrive; downwards a few thousand meters in oceans, beyond the reach of sunlight, where teeming populations of living things exist; and beneath the land, where bacteria have been found in petroleum reservoirs around 2000 m below the surface. Within this narrow band above, at, and below sea and ground levels, living organisms play vital roles in cycling and dispersing dozens of chemical elements.

VITAL ROLE OF ELEMENTAL REPLENISHMENT
IN THE BIOSPHERE

The stores of chemical elements within reach of organisms in the biosphere may appear to us to be large, but they are nowhere near big enough to sustain life without continuous replenishment. During the immense length of time life has existed on Earth, vast numbers of living things have lived, died, and been replaced by successive generations. No store of chemical elements, no matter how large, could have sustained life for so long without recycling. This is why the geochemical and biogeochemical cycles are so vital for life on Earth.

CYCLES OF FIVE CHEMICAL ELEMENTS CENTRAL TO LIFE

Of the 30 or so chemical elements essential for life, five – **hydrogen, oxygen, carbon, nitrogen**, and **sulfur** – stand out because they circulate, not just as solids along with all other minerals in rocky and sedimentary materials, as mentioned above, but variously as solids, liquids, and gases. These five also happen to be particularly central to life as we know it. The first two form water but all five may also, in an array of chemical combinations, dissolve or suspend in water and flow to the ocean where they may eventually sink to bottom sediments, not resurfacing for eons. On the other hand, unlike other elements, all five can also form stable gaseous compounds which cycle rapidly and continuously through the atmosphere. Whichever way, their cycling and dispersal are intimately linked to plants.

WATER CYCLING

The water cycle, the most rapid of all and the most massive on a global scale, starts with evaporation into the air, followed by transport on wind currents, precipitation as rain, snow, or fog, and flow down rivers. Dominated by the vast store of water in the oceans, it is controlled by energy from the Sun.

Over 97% of all water on Earth is found in its oceans, which occupy around 70% of the Earth's surface. Together, these realities give to the oceans a controlling influence over the Earth's climate from the tropics to the poles. About half the energy from the Sun which reaches the Earth's surface is absorbed by the oceans and then transferred to the atmosphere by evaporation of water. Some of this heat energy may be transported thousands of kilometers before being released when water vapor in the air falls as rain (water absorbs heat as it changes from liquid to vapor, releasing it again when the reverse occurs). Hurricanes, typhoons, monsoons, cyclones, and lesser storms play an essential role in moving moisture and heat energy from tropical waters to higher latitudes; the major ocean currents also transfer massive quantities of heat energy across the planet. Together, these natural processes ensure that many places that might otherwise be too cold or dry can support populations of organisms.

The amount stored in living tissues is far less than 1% of all water on Earth even though organisms are made up of between 60 and 90% water. Still, plants play a significant role in water retention and cycling both regionally, in particular wherever forests dominate, and globally, since they contribute, through evapotranspiration, an estimated 10% of all water entering the atmosphere.

CARBON CYCLING

Rapid cycling through photosynthesis

The most familiar carbon cycle, and the most rapid, is through plants and certain bacteria taking CO_2 from the air in photosynthesis, releasing it again to the atmosphere when they die and are decomposed. Photosynthetic organisms withdraw an estimated 100 billion tonnes or more of carbon as CO_2 from the atmosphere every year. This rapid cycling regulates the content of CO_2 in the air on a timescale of one or two to perhaps a few thousand years depending on the rate of decomposition of dead plant remains in different regions of the globe (Figure 15).

The initial understanding of this fast cycling of carbon by plants, and, as we shall see later, oxygen, nitrogen, and sulfur as well, within the biosphere depended on the discovery of the role microorganisms play.

Biome	Global carbon stocks (% of total)
Forests	
▫ Boreal	24
▫ Tropical	17
▫ Temperate	6
Grasslands	
▫ Tropical savannahs	13
▫ Temperate grasslands	12
Wetlands	10
Deserts, semideserts	8
Tundra	5
Croplands	5

FIGURE 15 Proportions of total global carbon stocks in soil and vegetation in the major biomes of the Earth. The estimated total amount of carbon: 2.3 Teratonnes (Tt; teratonne = 10^{12} tonnes) (after Cowie, 2007).

Antonie van Leeuwenhoek

Microorganisms were revealed to the world during the last years of the seventeenth century through the remarkable skill of **Antonie van Leeuwenhoek**, a Delft draper, whose hobbies included glass lens-making. His were the most perfect and powerful lenses of his age, allowing him to built the most successful microscopes then available. With them, he uncovered a whole new universe of, what he called, *animalcules* (little creatures), including bacteria, living in such diverse places as rain and seawater, on the surface of seeds, even in his mouth.

Charles Cagniard-Latour and Theodor Schwann

Leeuwenhoek's animalcules remained nothing more than curiosities for over a century until about the 1830s when the French inventor, **Charles Cagniard-Latour**, and the German physiologist, **Theodor Schwann**, discovered a correlation between alcoholic fermentation (see Chapter 2) and the presence of the microorganism, yeast. They independently observed that grape juice became wine only when yeast cells were growing there. This pioneering conclusion led others later, notably **Louis Pasteur**, to establish the many critical roles that microorganisms (bacteria, yeasts, and other fungi) play in the biosphere.

We now understand that the decomposition of dead remains of organisms, notably the huge amounts of plant debris generated in the biosphere, is carried out by a vast array of microorganisms (mainly fungi in the case of plants, bacteria in the case of animal remains), which, through their respiratory activities, make carbon as CO_2 once again available in the atmosphere for capture in photosynthesis.

Long-term carbon cycling

Long-term carbon cycling is entirely different from and dwarfs its more rapid, short-term counterpart. This hugely important, stately process cycles hundreds of times as much carbon as is contained in the world's terrestrial biosphere every million years or so.

Long-term cycling of carbon begins with the eruption of volcanoes, which vents CO_2 from deep in the Earth into the air. Once in the air, CO_2 dissolves in rainwater, forming carbonic acid (H_2CO_3), which, although weak, will slowly break down silicate and carbonate rocks, flushing trapped ions, including, importantly, calcium (Ca^{2+}), into rivers. In aqueous solution H_2CO_3 is in equilibrium with the CO_2 from which it is formed and is also ionized to H^+ and bicarbonate (HCO_3^-); the latter, in turn, ionizes to H^+ and carbonate (CO_3^{2-}). The relative amounts of these components depends on the pH of the water in which they are dissolved.

Incidentally, plants themselves also accelerate the acid breakdown of rocks. Plant roots secrete into soil small quantities of organic acids which speed up about five-fold the rate at which rocky materials undergo chemical weathering. Even plant litter at the soil surface creates an acidic environment within which soil minerals are dissolved into soil water solution more rapidly.

Once in the ocean, marine organisms use the CO_2 from bicarbonate for photosynthesis and the calcium flushed from rocks, together with carbonate, to construct their shells. When these organisms die, their shells drift down, adding to sediments accumulating on the ocean floor. The well known Dolomite mountains in Italy and the White Cliffs of Dover in England are largely made up of calcium carbonate shells of marine organisms uplifted from ancient sea beds. Sooner or later some of these sediments are carried deep into the Earth's hot mantle where they are cooked at sufficiently high temperature and pressure to release their CO_2 again to be vented through volcanoes.

Crucial role of carbon dioxide in the atmosphere

This recycling, driven by the fierce heat in the Earth's interior, is crucial to maintaining some level of CO_2 in our atmosphere, which, in turn, acts to keep temperatures within the limits between which life can flourish on our planet. Carbon dioxide may be, to us, an alarming greenhouse gas but, if there were none of it in our atmosphere, the Earth would be a cold, barren wasteland.

OXYGEN CYCLING AND CYANOBACTERIA

The evolution of photosynthesis was the major factor in the appearance of oxygen (O_2) in our atmosphere. Today, the element makes up about a fifth (21%) of the gases in air but this was not always so. During the first billion years or so of the Earth's existence there was no detectable oxygen in the atmosphere. The earliest life forms, primitive microorganisms, were adapted to a reducing atmosphere rich in nitrogen (ammonia; NH_3), CO_2 and sulfur (hydrogen sulfide; H_2S) gases which poured out from volcanoes. The appearance of blue–green **cyanobacteria** about 3.5 billion years ago eventually led to dramatic changes in the atmosphere.

Oxygen, a byproduct of the ability of cyanobacteria to harness light from the Sun through photosynthesis as a source of energy, led to a slow but highly significant change in the Earth's atmosphere towards that present today, although it remained a trace element in air for about another billion years. The O_2 from the earliest photosynthetic organisms was, at first, trapped in the hydrosphere by ferrous iron (Fe^{2+}), producing ferric iron (Fe^{3+}). The latter precipitated out of solution to give rise to **banded iron formations** (BIFs; Fe_2O_3), our iron ore deposits. These stopped accumulating about 2 billion years BP when all the Fe^{2+} in solution was used up.

As O_2 then began to accumulate in the atmosphere, conditions became favorable for the evolution of aerobic eukaryotes (e.g. algae), the abundance of which remained limited in surface waters, however, until the ozone layer in the stratosphere developed to protect against UV radiation from the sun. As the concentration of O_2 in the atmosphere rose, the way was opened for the appearance of green plants (about 400 million years BP), an evolutionary development which, in turn, accelerated the rate of oxygen accumulation in the air.

Why oxygen accumulated

Although we know that it is *released* during photosynthesis, we also now know that O_2 is *removed* from the atmosphere in exactly equal amounts during respiration, a process common to all organisms (see Chapters 1 and 2). Thus, you would expect that for every molecule of O_2 added to the air by a plant during its lifetime, exactly the same number of molecules would be removed during the breakdown of plant bodies when they die and are decomposed by microorganisms, resulting in no net gain to the atmosphere.

However, when plants die they are not always completely decomposed. A small fraction of their biomass, less than 1%, remains as debris on land or as suspended organic matter in water. We see evidence of this in wetlands such as bogs and the Arctic tundra where vast peat beds have accumulated. Huge quantities of organic matter carried by rivers are discharged into coastal waters, giving rise to massive mud flats. The Amazon, for example, by far the largest river system on Earth, carries to the coast every year an estimated 70 million tonnes of carbon locked in the remains of, mostly, plants. A fraction of this material forms ever thicker deposits of organic matter, some of which sink to the ocean floor as sediments and end up deep in the Earth's mantle where they may remain for eons. Our coal deposits are one result of these actions.

The crucial point is that not all the O_2 released into the air during photosynthesis is removed later when plants are broken down because a small fraction of the original plant material formed through photosynthesis is not completely consumed. The outcome is that the atmosphere has had added to it small fractions of unused O_2 over hundreds of millions of years, giving us the 21% we have today. Wetlands (mud flats, bogs, tundra) are sometimes referred to as the "lungs" of the Earth, which, on a geological timescale, result in a slow exhalation of O_2 into the atmosphere.

Eventual recycling ...

However, eventually, stores of organic material, too, will be consumed. It may take an unimaginably long time but the cycling of the Earth's matter never ceases. Unconsumed plant debris may remain close to the surface (e.g. peat) or be compressed into sedimentary rock and sink deep into the Earth's crust (e.g. coal), from where they will be uplifted to the surface many millions of years later. Once brought in contact with O_2 in the atmosphere, these organic remains, weathered by wind and rain, are oxidized back to the carbon dioxide and water used in their original formation by photosynthesis millenia before.

... Accelerated by human activity

We are currently speeding up this slow, balanced process immensely by burning the fossil fuels of the Earth. Along with coal, oil, and natural gas reserves are other results of the incomplete breakdown of organic matter stored deep in the Earth's crust. Our acceleration of the release of CO_2 from these reserves is overwhelming the much slower natural process, leading to unnatural, rapid changes in global climate. There is also concern that climate change will lead to a melting of the Arctic tundra, releasing the massive amounts of CO_2 and methane (CH_4; which is over 20 times more effective than CO_2 as a greenhouse gas) known to be stored there. The release of CH_4 from methane hydrates on the edges of continental shelves into the atmosphere is thought to have been important to global warming in past geological eras.

On the bright side, the store of O_2 in the atmosphere, which is very large, is in no danger of being depleted by our burning of the fraction of fossil fuels we can reach and extract economically.

NITROGEN AND SULFUR CYCLES

Microorganisms also cause the release of nitrogen and sulfur from the organic matter they break down. The CO_2 they return to the air can re-enter the biosphere directly through photosynthesis but how do nitrogen and sulfur circulate?

NITROGEN CYCLING

About 97% of all nitrogen is in the crust, mantle, and core of the Earth; most of the rest, about 2.3%, makes up 78% by volume of the gases in our atmosphere. Nitrogen is brought to the surface from deep in the Earth either through volcanic action, as ammonia gas, or by geothermal up-welling of water, which often brings with it high concentrations of dissolved ammonium ions (NH_4^+). The nitrogen vented in gaseous or dissolved form from deep in the Earth is thought to have been in the mantle for billions of years and to be a possible remnant of the Earth's early atmosphere.

The planet's nitrogen cycle changed dramatically once photosynthetic organisms evolved. It was then that the highly corrosive O_2 molecule began to appear at the Earth's surface, changing everything, albeit slowly. Once exposed to oxygen, the highly reduced sources of nitrogen, primarily ammonia, were oxidized either to nitrogen gas (N_2) in the atmosphere or nitrate (NO_3^-) in soil and water.

Scarcity of nitrogen in the atmosphere and biosphere

The main input of nitrogen into the living world comes from the limited reservoir in the atmosphere that only certain bacteria can access; few natural processes have evolved which can break the strong triple bond ($N\equiv N$) of the nitrogen molecule (see Chapter 5). The rest of the Earth's nitrogen is locked firmly out of reach of the biosphere in igneous and sedimentary rocks from which it is released only very slowly.

Thus, although nitrogen is essential for all forms of life it is found in the biosphere in only minuscule amounts (about 0.0002%) of the total nitrogen on Earth. Also, it is not very soluble in water except in its inorganic forms, ammonium (NH_4^+), nitrite (NO_2^-) and nitrate (NO_3^-). Living and dead organic matter in soil is an additional reservoir of nitrogen available to the biosphere but, again, amounts are small. This relative scarcity of nitrogen within the biosphere is important to the distribution of life on the planet.

The element is particularly scarce in a number of terrestrial ecosystems, such as many grassland, boreal forest, and montane communities. There are two major reasons for this: the element is easily lost through the leaching of highly soluble nitrates in the soil down below the reach of roots and by the evaporation of ammonia from soil. The element's low concentration in water also limits the productivity of many aquatic species. Even the nitrogen already incorporated into living tissues is released from decomposing organic remains only slowly and with difficulty.

The dominance of bacteria in nitrogen cycling

Thus, the cycling of nitrogen in the biosphere stands apart from others by being largely dominated by bacteria.

Organic nitrogen begins its transformation to inorganic forms with the decomposition of the large molecules found in dead organisms, such as proteins, RNA and DNA, by a bewildering array of microorganisms in the soil. This is a slow process, the microorganisms managing to release in a year only about 1% of the total organic nitrogen stored in soils. In some soils, such as those in cold boreal coniferous forests, plant litter may remain for centuries before being broken down. Recycling is somewhat accelerated where there are large herds of grazers, such as deer, elk, Arctic caribou, ungulates of the African savannahs, and the historic bison of the North American plains, which pass nitrogen through their digestive systems, out into the soil.

Cycling of nitrogen in soil

The cycling of nitrogen in soil begins with the transformation of the ammonia released from organic matter to ammonium ions (NH_4^+). Fortunately, this highly toxic ion is either quickly oxidized to nitrate by **nitrifying bacteria** or removed from rapid circulation altogether by adsorption onto clay minerals. Nitrate is the favored form of nitrogen by plants and is readily absorbed by them, but it also disappears from soil quickly; because of its high solubility in water it may be rapidly washed down into ground water beyond the reach of plant roots. Alternatively, it can be converted to N_2 gas by **denitrifying bacteria** or reduced back to NH_3 gas by **fermentative bacteria**, both of which gases may end up in the atmosphere.

Following the fate of an atom of nitrogen in soil is complex. It may move from soil to microorganism and back to soil in less than a day; soil to plant to soil over a period of months; end up adsorbed by clay for decades before being released again through weathering; held as humic acid in peat and other long-term deposits of plant debris for millenia. Thus, even soils with an apparent large nitrogen content may have less than 1% of it available in a form that plants can access at any moment in time. Not surprisingly, plants have found ways to scavenge and conserve nitrogen other than through a dependence on soil microorganisms.

Alternative ways plants have to recycle nitrogen

In some soils where nitrogen is in particularly short supply, plants absorb as much as possible of it in organic form. For instance, dead organic matter releases **amino acids** into the soil, from the breakdown of proteins, which can then be taken up directly by nearby plants. Alternatively, many plants recycle nitrogen internally, not allowing it to escape at all; as they prepare to shed their leaves before the onset of winter or when they die back after flowering, they move essential minerals, including nitrogen, into surviving stems and roots where they are stored until needed for renewed growth.

In areas where destruction of plants and plant debris by fire is a major factor, the amount of nitrogen released into the air and soil can be as high as one-half of all that withdrawn from the atmosphere by nitrogen fixation in that region (see Chapter 5).

SULFUR CYCLING

Among the gases vented through volcanoes in the early Earth were **sulfur dioxide** (SO_2) and **hydrogen sulfide** (H_2S), neither of which remained in the atmosphere for long because of their high solubility in water. Once the oceans formed, sulfur compounds were quickly removed from the air: SO_2 reacts with water to form sulfate (SO_4^{2-}; by far the most abundant form of sulfur in sea water), while H_2S gives rise to iron pyrites (FeS_2; fool's gold) by reacting with oxides of iron, which are ubiquitous in soil and other rocky material. But, as in the case of other chemical elements we have discussed, the appearance of life on Earth profoundly altered the physical cycling of sulfur.

Photosynthetic sulfur bacteria

The evolution of cyanobacteria led to a photosynthesis in which water is used to reduce CO_2 to sugars, releasing oxygen, an *aerobic* process which dominates the green world today (see Chapter 1). But, before the emergence of this process, certain bacteria had evolved that were capable of using either sulfur itself or H_2S rather than water as reductants in an *anaerobic* form of photosynthesis. Even today, descendants of these early life forms, the **green and purple sulfur bacteria**, use this primordial process, releasing the highly oxidized sulfate ion, not oxygen, as a byproduct of their reduction of CO_2 to carbohydrate (see Chapter 1, Box 1).

Short cycling in the soil

In addition to the sulfate produced by photosynthetic sulfur bacteria, a number of other paths end with this most highly oxidized form of the element.

A mixture of sulfur compounds is released on land or at its edges by the action of an array of microorganisms. The most abundant of these compounds, H_2S, is released from soils, coastal marshes, mud flats, mangrove swamps, freshwater wetlands, and at the bottoms of lakes, wherever oxygen is scarce or absent. Eventually, all these sources of sulfur compounds are oxidized to sulfate, which plants then take in and use as their main source of sulfur. They promptly reduce it to H_2S again and incorporate it into such molecules as the amino acids, cysteine and methionine, and vitamin B_1, thiamine. The sulfur returns to the soil during decomposition of biomass, ending up once more as sulfate.

This tight cycle operates independently of any other source of sulfur. However, there is one major source from which sulfur can be diverted into a long-term storage reservoir: iron pyrites.

Long-term storage and cycling

Inside coastal mud flats, certain bacteria, the awkwardly named **thiopneutes**, reduce SO_4^{2-} in sea water to H_2S, which quickly reacts with iron oxides in the mud, producing iron pyrites (FeS_2). The ancestors of the modern thiopneutes were among the earliest life forms. About 95% of the sulfur found today in the crust of the Earth has an origin in the activity of these early bacteria in coastal waters. The burial of FeS_2 along with **evaporites** (precipitates of sulfate minerals formed directly from evaporation of sea water) deep in the sediments of the Earth removes huge amounts of sulfur to a reservoir far beyond the reach of living organisms. Reserves such as these are only available again to the biosphere after tectonic uplift and weathering or volcanic venting, a cycling that operates over tens or hundreds of millions of years.

Dimethylsulfide

Of course, the large reservoir of sulfur in the oceans is recycled back to land to be available once again as a nutrient. H_2S oxidizes too quickly for this role. However, another gas, **dimethylsulfide (DMS)**, is more stable than H_2S and is released from a number of marine algae, including, notably, the **haplophytes**, the organisms responsible for the chalk deposits of the White Cliffs of Dover in England. DMS entering the atmosphere from these sources becomes incorporated into clouds, eventually returning to the land dissolved in rain, snow, or fog.

SUMMARY

- Geochemical cycling of matter has been present from the earliest days of the Earth's existence
- Biogeochemical cycling was added once life evolved
- Over billions of years, these two types of cycles have determined, sustained, and regulated the rate and direction of changes to the Earth's physical and biological environments
- Both are crucial to the wellbeing of the biosphere, the domain of living things
- Cycles particularly central to life are those of the five elements that cycle in solid, liquid, and gaseous forms in the lithosphere, hydrosphere, and atmosphere of the globe – namely, hydrogen, carbon, oxygen, nitrogen, and sulfur
- These elements happen to be the five most abundant components of all living organisms. Maintaining a balance in their cycling and recycling is, therefore, important for the health of the biosphere as a whole
- At least from the start of the Industrial Revolution onwards, these cycles have been, and continue to be, increasingly distorted by ever-growing human populations and their activities
- Some consequences for plants of this burgeoning human intervention in the cycling of matter are the subject of Chapter 16.

16 The human touch

INTRODUCTION

The focus of the previous chapter was on factors that shaped and directed changes to the Earth's environment over the four and a half billion years before the start of the Industrial Revolution in the eighteenth century. But what effects have humans had since global industrialization began? How are the changes to the global environment caused by humans since then influencing plants, in particular, today?

Answers to these and other key questions are by no means complete. Even where answers have been provided, the data on which they are based and their interpretation are often contradictory or contentious. This is not surprising but is confusing even for those who have the broadest and deepest knowledge. Not only is human influence growing and changing constantly but the environment itself is composed of a complex array of components which interact with one another often in ways about which there is inadequate knowledge or none at all.

THE IMPACT OF HUMAN ACTIVITIES ON THE ATMOSPHERE

Central to any understanding of human influence on the global environment is our effect on the atmosphere. The composition of the Earth's atmosphere determines inward transmission of the Sun's energy, its distribution across the globe, and its radiation back into space. The difference between the incoming and outgoing energy from the Sun determines the surface temperature of the Earth, important to organisms because all have a temperature range to which they are adapted.

SVANTE ARRHENIUS AND THE FIRST GLOBAL CLIMATE MODEL

The carbon cycle in particular plays a key role in determining how much of the Sun's energy reaches the Earth's surface, is distributed across it, and escapes back into space.

Svante Arrhenius, a nineteenth-century Swedish chemist and Nobel laureate, was the first to propose a global climate model recognizing CO_2 as the dominant greenhouse gas. He concluded that a geometric increase in CO_2 in the atmosphere would lead to a near arithmetic rise in temperature at the Earth's surface and that the temperature would rise most in polar regions. He predicted that a doubling of CO_2 concentration would produce a warming of 5–6°C, a conclusion comparable to current climate models of the atmosphere (1.5–4.5°C).

GREENHOUSE EFFECT: EVER PRESENT, ESSENTIAL FOR LIFE

Before going further we must re-emphasize a point made in the previous chapter: Some greenhouse effect is natural, a consequence of the Earth having an atmosphere, and is essential for life. Our moon has no atmosphere; its day-time temperature is a searing +120°C; at night, a numbing –170°C; the median, –28°C. Earth's median, by contrast, is a balmy +15°C.

The warming effect on our atmosphere has always been there. Carbon dioxide, water vapor, methane, and oxides of nitrogen are all greenhouse gases and have all likely been present in our atmosphere for most of its existence (see Chapter 15). All allow the Sun's radiation to penetrate to the Earth's surface but prevent some of it escaping back into space. There have been times in the Earth's history when one or more of these gases were present at much higher levels than they are now, causing the median temperature of the Earth to be well above its present value.

CARBON DIOXIDE AND RADIATIVE FORCING

At the center of recent and ongoing concern is the extent to which all or some of these four gases are increasing in concentration in our atmosphere because of human activities and, principally, to what extent the global warming taking place now has its roots in increasing CO_2. Climate scientists use the concept of **radiative forcing** to quantify the effects of increasing levels of particular gases on climate and define it as the change in global energy balance compared to pre-industrial times. Positive forcing induces warming; a negative one, cooling.

PRE- AND POST-INDUSTRIAL LEVELS OF CARBON

Just before the Industrial Revolution began, the CO_2 level in the Earth's atmosphere was about 280 parts per million (ppm); now, in 2010, it is about 385 ppm and rising at about 2 ppm per year. Analyses of ancient air trapped in bubbles in Antarctic ice cores show that the level of CO_2 is at its highest level in more than half a million years. The years since the advent of industrialization have seen an alarming, rapid rise of 36% in CO_2, 150% in methane, and 17% in nitrogen oxides. The current amount of carbon from all sources in the atmosphere is estimated to be over 3.6 trillion tonnes and to be increasing at about 2 billion tonnes a month.

The question is, how much of these increases is due to humans and how much to natural fluctuations which we know have always been a feature of our planet's evolution (Figure 16).

HUMAN INFLUENCE ON GREENHOUSE GASES

International Panel on Climate Change (IPCC)
There is growing confidence that most, if not all, current increase in greenhouse gases is caused by human activities. Much of this confidence is based on results compiled by the Intergovernmental Panel on Climate Change (IPCC), a consortium of thousands of scientists from around the world who contribute information and data to a common pool from which assessment reports are compiled. A sample of the conclusions from these data published in the Fourth IPCC Assessment Report (2007) can be found in Box 16.

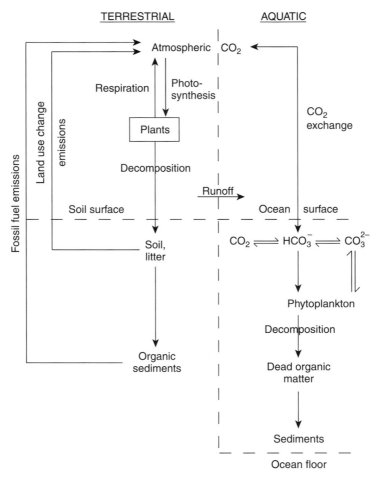

FIGURE 16 The carbon cycle today. The main human influences are through the burning of fossil fuels and changes to the use of land for agriculture (after Smil, 2001).

BOX 16. THE INTERGOVERNMENTAL PANEL ON
CLIMATE CHANGE (IPCC)

The IPCC (www.ipcc.ch) is a scientific body established in 1988 by the United Nations Organization to evaluate the risk of climate change caused by human activities. So far, the IPCC has published four reports, the latest in 2007; others will follow. The

BOX 16. (cont.)

Panel includes scientists from around the world. Its assessments are based mainly on peer-reviewed and published scientific literature, the product of investigations by thousands of researchers in all relevant science communities, worldwide.

Among conclusions in the Fourth Assessment Report (2007) are that:

- *Eleven of the 12 years from 1995 to 2006 were among the 12 warmest since 1850*
- *There is a better than 90% probability that increases in average temperatures since the mid-twentieth century are due to greenhouse gas emissions generated by human activities*
- *The oceans have absorbed about 80% of the heat added to the climate system and have warmed to a depth of 3000 m*
- *Sea levels rose at an average rate of 1.8 mm year^{-1} over 1961–2003 but by an average of 3.1 mm year^{-1} from 1993 to 2003*
- *Average Arctic temperatures increased at about twice the global average rate in the past 100 years*
- *The content of CO_2 in the atmosphere in 2005 (379 ppm) exceeded by far the natural range over the past 650 000 years (180–300 ppm), as determined from ice cores*
- *The primary source of the increased concentration of CO_2 in the atmosphere since the pre-industrial period results from fossil fuel use with changes to land use being another, smaller, source*
- *Methane in the atmosphere has increased since pre-industrial times from 715 ppb to 1774 ppb in 2005*
- *The combined radiative forcing owing to increases in CO_2, methane, and nitrous oxide is +2.30 W m^{-2}. There is a better than 90% probability that the rate of increase in radiative forcing during the industrial era is unprecedented in more than 10 000 years.*

From these conclusions, the IPCC predicts that, by the end of the twenty-first century, climate change will result in, among others:

- *A probable temperature rise between 1.8°C and 4°C*
- *A sea level rise between 28 cm and 43 cm*

BOX 16. (cont.)

• *The disappearance of Arctic summer sea ice in the second half of the century.*

Not everyone agrees with the IPCC conclusions and predictions. For example, some detailed counter-arguments from and for skeptics can be found at www.climatechangefacts.info *among many other web sites.*

Changes causing cooling

Not all human activities lead to warming of the planet. **Sulfate aerosols**, which give rise to **smog**, are also caused by fossil fuel combustion and biomass burning. These have a cooling effect by preventing sunlight reaching the Earth's surface. The haze which forms over major populated areas, such as North America, western Europe, and east Asia, especially during the summer months, cools the globe. **Sulfuric acid** (H_2SO_4) and **ammonium sulfate** ((NH_4)$_2SO_4$) are the most abundant aerosols resulting from human activities. Together, they lower the temperature of the troposphere (the lowest portion of the Earth's atmosphere) by scattering some incoming energy from the sun back into space. Estimates from the 1990s suggested a cut of about 0.3% (0.5 W m^{-2}) in incoming energy; this compared to an estimated gain from a higher concentration of greenhouse gases in that same time period of about 2.3 W m^{-2}.

Thus, the cooling effects of sulfates may be one possible answer to the puzzle as to why the warming of the planet since the end of the nineteenth century has been less than predicted from climate models. The global rise in temperature has been between about 0.6°C and 0.8°C since then; the rise of CO_2 and other greenhouse gases during the same period should have translated to a rise of up to 1.3°C, it is estimated.

Water vapor is also a powerful greenhouse gas, contributing significantly to the natural, as opposed to the human-induced, greenhouse effect. Warming increases evaporation of water from the Earth's surface into the troposphere, adding further positive forcing of temperature. On the other hand, increased water vapor leads to increased cloud formation; clouds reflect sunlight back into space, cooling the Earth's surface during the day while acting as a greenhouse blanket, warming it at night (the **aerosol cloud albedo effect**).

Water vapor and sulfates, where present together, contribute yet further to the aerosol cloud albedo effect. Airborne sulfates attract water very efficiently, creating nuclei around which water droplets coalesce into clouds.

Finally, **dust** from volcanic eruptions and agricultural practices is known to act as a significant atmospheric coolant, periodically and transiently. It has even been suggested that the short dips in temperatures during the early 1940s were caused by dust from World War 2 actions.

Correlation of carbon dioxide and global temperature

Regardless of complications as a result of smog or other modifiers, over the period from 1958 to 1989, CO_2 concentrations in the atmosphere and global temperatures have been correlated to a confidence level greater than 99.9%. This suggests strongly that the contribution of human forcing to the carbon cycle is by far the most significant component of climate change.

PLANTS AND CLIMATE CHANGE

Since plants are major components of the biosphere, it is important to understand how their daily lives are affected by changes to climate brought about by human activities. Some examples are provided here to illustrate how plants are reacting, physiologically, to changes to the global environment and how these responses may impact, in turn, on climate change and the cycling of some nutrients important to the entire biosphere.

PLANTS AND CARBON DIOXIDE

CARBON DIOXIDE AND PHOTOSYNTHESIS

Direct effect

Total global terrestrial photosynthesis has been estimated to remove about 20-fold more carbon as CO_2 from the atmosphere than is added to it by fossil fuel combustion. This sounds promising as a way to lessen global warming except that respiration by all organisms, including plants, also adds CO_2 to the air. As long as photosynthesis outpaces all sources of additions of the gas to the atmosphere, the biosphere will act as a carbon sink. Clearly, this is not the case today since the concentration of CO_2 in air is increasing; photosynthesis is not keeping pace.

Only a few years ago, experimental results on photosynthesis seemed to provide hope that C_3 plants (see Chapter 1) would accelerate their photosynthesis in response to increasing levels of atmospheric CO_2. Results from short-term studies under laboratory conditions showed dramatic increases in C_3 plant growth at high CO_2 levels. Such results suggested that ever higher levels of CO_2 might have a positive influence on C_3 plants, leading to a greening of the Earth producing, in turn, an endlessly increasing capacity to mop up the greenhouse gas as fossil fuel consumption grew. Alas, longer term measurements showed a rapid decay of this initial growth spurt, but why?

There may be more than one answer to this question, one of the most important being a downshift in photorespiration (see Chapter 2).

Indirect effect

As the level of CO_2 in air rises, C_3 plants direct it more efficiently into photosynthesis, away from photorespiration. In the short term this stimulates growth of the plant as more food is manufactured.

But diverting CO_2 away from photosynthesis is not the only role of photorespiration. It also contributes to the nitrogen status of

C_3 plants by aiding in the synthesis of the amino acids, glycine and serine, and in the reduction to ammonia of nitrate taken up by plant roots from the soil. Anything that interferes with a plant's ability to take in and process nitrogen *lowers* growth by slowing the formation of vital molecules like DNA, RNA, and proteins.

It's all about the nitrogen

Nitrogen is often in short supply in the biosphere (see Chapter 5); thus, a plant must apportion it among its many competing sinks (see Chapter 4). Higher demand for it to increase photosynthetic capacity because of higher CO_2 levels needs to be balanced by the plant with accelerated growth of shoots, roots, leaves, flowers, and fruits in response to the increase production of food in photosynthesis. The outcome of this competition is a slowing of photosynthesis in favor of maintaining balanced growth.

Thus, although in the short term higher CO_2 accelerates carbon cycling in the biosphere directly by stimulating photosynthesis, in the longer term, the effects of elevated CO_2 become more indirect. It may take several years but, eventually, other components of the environment come into play, such as the availability of nutrients, including nitrogen, which may be there in sufficient amounts for a surge in photosynthesis to begin with as CO_2 rises, but which cannot sustain a growth increase in the long term.

Liebig's "Law of the Minimum"

The controlling influence of nitrogen on growth illustrated here is a classic example of the "Law of the Minimum" in action, a concept popularized by Justus von Liebig (see Chapter 4).

The lesson learned

The lesson learned here is that short-term changes to photosynthesis in response to higher CO_2 levels are not a reliable predictor of how plant growth will change in the long term and, thus, how the carbon cycle will be affected over time.

C_4 plants unaffected

It should be noted that C_4 plants remain largely unaffected by increases in CO_2; they have no photorespiration and, therefore, little spare capacity for increased rates of photosynthesis as CO_2 levels rise.

CO_2 AND PLANT WATER RELATIONS

Another major influence on plants is the availability and distribution of water in the environment in response to climate change.

Increasing CO_2 has its main effect on plant water relations through its influence at the boundary between the outside and inside of the plant, i.e. at the leaf surface. At this junction are tiny pores, the stomata, through which gases of all kinds enter and exit the plant (see Chapter 3). It is here that CO_2, especially, but also ozone as we shall see later, exerts key influences on the movement of all gases into and out of the plant, including water vapor.

Effect on stomatal aperture, transpiration, and run-off

Plants respond to higher CO_2 concentrations by partially closing their stomata. In this way, they can continue to take in as much CO_2 as they need for photosynthesis yet, at the same time, slow down their loss of water in transpiration (see Chapters 3 and 12). But what are the consequences of this to the cycling of water in the environment and to plants themselves?

Lowering transpiration also lowers the amount of water that plants need to take up from the soil through their roots to maintain an optimum water status. If plants absorb less water, more of it remains in the soil, which, in turn, increases run-off into rivers.

Recent experiments in which plants were kept in an atmosphere where CO_2 concentration was doubled showed that the mean run-off from natural terrestrial ecosystems has increased by an estimated 6% since pre-industrial times. Studies with the crop plant, soybean, at two different CO_2 concentrations reached a comparable conclusion; in this case, there was a decrease of 8.6% in the evapotranspiration of water vapor from stomata into the surrounding air at the higher CO_2 concentration.

What effect do these results have on how to estimate the influence of climate change on the water cycle?

Conclusion – physiological forcing is also important

Both of these studies, one natural, the other agricultural, suggest that, while higher CO_2 is causing the temperature of the globe to increase, leading to a higher evaporation of water from the Earth's surface, such a physical forcing of the water cycle is not the whole story. Increased run-off owing to decreased use of available water by plants (**physiological forcing**) must also be taken into account. Consideration of direct temperature forcing alone will tend to underestimate future increases in run-off and overestimate decreases.

Effect on stomatal density

Another example of physiological forcing is the apparent ability of some, maybe all, plants not only to vary the size of openings in their stomata in response to fluctuating CO_2 concentrations but also to vary their density.

The number of stomata per unit area in leaves of eight species collected between 1787 and 1987 in Britain and stored in herbaria was shown to have decreased by 40% over the 200-year period. The decline in stomatal density was correlated with an increase of 21% in the concentration of CO_2 in the atmosphere.

A causal connection between the decline in stomatal density and rise in atmospheric CO_2 was established in parallel experiments in which living examples of these same species were grown under controlled environmental conditions at varying concentrations of atmospheric CO_2. In the most extreme case a change in CO_2 concentration from a pre-industrial level of 280 ppm to the 1987 level of 340 ppm caused a decrease in stomatal density of 67% in new leaves formed during a 3-week period.

So strong is the association between stomatal density and CO_2 that geologists use the number of stomata per unit area in fossil leaves to estimate the concentration of CO_2 in the atmosphere at the times these plants were living.

Long-term effects on water cycling

These results indicate that the influence of CO_2 on stomatal development in the long term is at least as significant as it is on stomatal aperture in the short term. Both will have a profound effect on water uptake and transpiration and, therefore, on the cycling of water within the biosphere. Many plants reduce growth in response to even small reductions in available water. If a plant takes up less water it also takes up fewer nutrients from the soil dissolved in the water; growth slows. Plant growth and development show a highly tuned sensitivity to changes in both soil and atmospheric water status.

OCEAN ACIDITY

Another crucially important effect of increasing CO_2 in the atmosphere is its influence on the oceans, which act to remove CO_2 and other greenhouse gases from the air. At first glance, this may seem advantageous; it might be expected that some of the CO_2 dissolved in ocean waters as bicarbonate (see Chapter 15) would be used to stimulate photosynthesis among phytoplankton (algae, diatoms, dinoflagellates, cyanobacteria), photosynthetic organisms which form the basis of marine food chains; not so.

Just as the shortage of nitrogen restrains increase in photosynthesis in land plants owing to a rise in CO_2 (see above), in the ocean, the main restraining factor is a chronic shortage of iron (Fe) in a form that phytoplankton can access. Iron is highly *in*soluble in sea water and what *is* in solution is mostly unavailable to aquatic organisms. Iron is an essential component of photosynthetic, respiratory, and nitrogen fixation pathways in cells; shortages of it, therefore, prevent significant growth spurts in response to increased CO_2 (another example of Liebig's Law of the Minimum in action).

Because of these growth restraints the billions of tonnes of CO_2 accumulating as dissolved gas in water since the start of the Industrial Revolution are causing a steady increase in ocean acidity. For most of the past 600 000 years, it is estimated that the average pH of the ocean was a stable 8.2; in the past 300 years or so this figure has dropped to 8.1, which may not seem much but pH is a logarithmic scale; this decline represents a 30% increase in acidity. The ocean is estimated to be absorbing man-made CO_2 at a rate of over 22 million tonnes a day and climbing, which, it is suggested, will give rise to a drop in pH to 7.8 by the end of the twenty-first century, a 150% increase in acidity since pre-industrial times.

This is a serious matter for organisms which depend on calcium carbonate for making shells, such as some marine algae (mainly the haplophytes), mollusks, and crustaceans, and for those organisms in marine food chains that depend on shellfish as a dietary staple. Consequences for marine phytoplankton and other plants, such as red and brown algae, are unknown at this point, but profound disruptions of food chains are rarely beneficial in the long term. The consequences for coral reefs will also be disastrous.

INFLUENCE OF OZONE ON PLANTS

Most **ozone** (O_3) is produced by the reaction of sunlight with O_2 in the stratosphere, where it acts to intercept ultraviolet light at 180–240-nm wavelengths. In this way, ozone protects the Earth's surface from that portion of the solar spectrum most damaging to living tissues.

The production of ozone in the troposphere, especially in the smog of polluted cities around the world, is the result of reactions involving NO_2 from fossil fuels and is a different matter. NO_2 is dissociated by sunlight into NO and O, followed by a reaction between O_2 and O to yield O_3. The level of this source of ozone in the troposphere is rising steadily as the burning of fossil fuels grows.

Tropospheric ozone causes more damage to plants than all other air pollutants combined. The gas enters leaves through stomata and, as a strong oxidant, causes several kinds of damage. Daily exposure to ozone leads, at first, to classic symptoms of stippling, flecking, bronzing, and reddening of leaves. As pollution grows in severity, these early symptoms are gradually overtaken by more widespread chlorosis and necrosis of leaves. All these symptoms interfere with the efficiency of leaves to photosynthezise to a greater or lesser extent and, thus, lessen growth in wild and cultivated plants.

Further, it is now recognized that continuing increases in O_3 and CO_2, together, may lead to a synergistic effect on stomata. As discussed above, increase in CO_2 causes partial closure of stomatal apertures but leaves can still import the amount of the gas they need for photosynthesis. However, it has now been demonstrated that O_3 can damage guard cells to the extent of causing stomata to remain closed. In this way, ozone contributes positive forcing to global warming by restricting the ability of plants across the globe to capture CO_2.

HUMAN INFLUENCE ON NITROGEN CYCLING

In the previous chapter as well as in this, mention was made a number of times that the scarcity of nitrogen in the biosphere places a significant restraint on how plants react to changes in their environment. But these changes also include nitrogen itself, which is being added to the atmosphere and water bodies, both fresh and marine, because of human activities. How are these forced changes affecting plants?

HISTORICAL INFLUENCE OF HUMANS ON NITROGEN AVAILABILITY

Since prehistoric times humans have used fire to modify their environment for their benefit through the deliberate clearing of land and the regular burning of waste agricultural biomass, leading to loss of soil and biomass nitrogen to the atmosphere. As agricultural practices became more sophisticated, increasing numbers of domesticated animals provided a source of slow-release organic nitrogen (manure) for use on crops. Industrialization shifted human influence on the nitrogen cycle yet again. Manure became increasingly replaced by chemical fertilizers and the planting of improved nitrogen-fixing legumes. Today, the main human activities influencing the nitrogen cycle around the globe are fossil fuel combustion, the production and use of chemical fertilizers, the growing of nitrogen-fixing crops, and human and animal waste disposal, all of which create environmental issues. It is estimated that the amount of nitrogen now in circulation in the biosphere is twice that of pre-industrial times.

FRESHWATER AND MARINE EUTROPHICATION

Freshwater pollution

Eutrophication is one important consequence linked to agricultural practices and waste disposal.

Freshwater plants gain the nutrients they need largely from the land round about by drainage. Chemical fertilizers dissolved in drainage water and fluid animal wastes can provide an overabundance of nourishment, notably nitrogen and phosphorus. These additives stimulate aquatic plants to increase their photosynthesis which leads, in turn, to unnatural, luxuriant growth, an example of which is the algal blooms commonly found in freshwater lakes and ponds around the world. When these overabundant plant biomasses die, their decomposition by microbes depletes the water of dissolved oxygen, resulting in widespread death of aquatic organisms.

Ocean pollution

Nitrogen fertilizer and human and domestic animal wastes carried to the ocean in water or on the wind also stimulate production of marine plant life, within the constraints imposed by the chronic shortage of iron in the ocean (see above), which can have both positive and negative consequences.

The increase in plant numbers means there is more photosynthesis for which more CO_2 is required. It is estimated that this higher photosynthetic activity in the ocean removes about 10% of the CO_2 added to the air because of human activities, thus potentially restraining global warming.

Unfortunately, not all the nitrogen added to the ocean leads to such a positive outcome. Some of that taken up by marine microorganisms is released back into the atmosphere as nitrous oxide (N_2O), a greenhouse gas about 300 times more powerful than CO_2, and which is estimated to cancel out roughly two-thirds of the gain from CO_2 removal from the atmosphere by marine plant life. In addition, marine eutrophication is leading to the increasingly worrying growth of massive **dead zones** in the oceans where the lack of oxygen eliminates all living things.

NITROGEN POLLUTION: AQUATIC AND TERRESTRIAL HABITATS

Results such as these strongly suggest that the addition of nitrogen to the ocean through the interference of humans in the nitrogen cycle is an important factor in climate change. For example, an estimated 55 million tonnes of nitrogen produced by human activities entered the ocean from the atmosphere in the year 2000, which led to re-emission of nitrogen gases to the atmosphere ten times greater than in the mid-nineteenth century.

During the present century, nitrogen pollution is expected to continue to rise and to have an increasing influence on terrestrial as well as aquatic habitats. Recent analyses of data from a large number of studies on plants from tropical forests to Arctic tundra show that excess nitrogen causes growth to increase in all ecosystems except deserts. Whether these effects will continue in the long term is uncertain. Already, there has been surprise at the average 20% boost to growth from excess nitrogen in tropical forests where there was thought to be ample nitrogen and where growth was expected to be constrained by low levels of phosphorus in soil, an expectation that may still prove to be the case over the long term.

Any time a natural level of a nutrient is doubled by human activity, as is the case for nitrogen, some negative consequences are to be expected. Nitrogen levels in the groundwater of some areas of lowland

Europe have risen, as a result of agricultural practices, to the extent that this source of water is no longer safe for consumption by babies.

PLANTS AND TEMPERATURE

As discussed earlier in this chapter, an important global influence of increasing atmospheric CO_2 concentration is on the temperature of air, land, and water bodies.

The temperature of plants varies with that of their environment; they are **poikilothermic** (poikilos = variable). Since they commonly cannot move, except when they form reproductive products like seeds that may be carried far away on the wind, in rivers, or on ocean currents, they must cope with whatever their environment offers; they must adapt or die.

What can be said about how plants adapt to higher temperatures, especially physiologically? Before providing examples to address this question, a couple of cases illustrate the importance of first assessing the influence of speed of change.

EFFECT AT HIGH LATITUDE AND ALTITUDE

Comprehensive, broad-based measurements globally have shown that at high latitude or altitude (i.e. low temperature environments near the poles and on mountain slopes), the amount of new plant biomass produced averages $0.4 \, kg \, m^{-2}$ year^{-1}, in a typical **2-month** growing season. In equatorial forests, on the other hand, annual productivity is about $2.5 \, kg \, m^{-2}$ year^{-1} in a **12-month** growing season. Surprisingly, the monthly productivity is about the same in both cases, around $0.2 \, kg \, m^{-2}$. Differences in annual productivity are related to time, not temperature: the mean temperature during the growing season is about 8°C in one case, 28°C in the other.

EFFECT IN THE TROPICS VERSUS THE ARCTIC

In another example, about one-half of CO_2 output from the soil is caused by the respiration of microbes. Since rate of respiration is sensitive to temperature, it might be expected that CO_2 release owing to the decomposition of plant litter by soil microbes would be far higher in the tropics than in the Arctic, but not so. The monthly rates of CO_2 evolution during the growing season are about the same. Annual CO_2 efflux depends on the total amount of litter available to soil microbes, not on temperature, again a function of time.

Conclusions from these examples

These two examples illustrate that even major differences in temperature may have no net effect on some key plant and ecosystem processes provided that changes occur slowly over a long period of time, allowing organisms, particularly microbes, to adapt. Both examples use data from natural systems where the vegetation and microbes have been in long-term equilibrium with one another and the resources available to them; they are fully adapted to their environments.

However, it is a different issue how plants and ecosystems as a whole respond to rapid change in temperature over a few short decades, as is the case now because of human forcing.

ECOSYSTEMS ARE NOT ALWAYS COMPARABLE

Also important to remember is that different ecosystems may have significantly different sources for the CO_2 they release. For example, as noted above, the source in forest communities is through decomposition of detritus; in deep aquatic habitats, on the other hand, most CO_2 is released through respiration. Thus, conclusions drawn from measurements of CO_2 release in forest communities may not apply to deep aquatic environments. Care must be taken in extrapolating from one ecosystem to others.

HOW RAPIDLY IS WARMING OCCURRING?

Current models of climate change predict that global temperature will increase at a rate of about 0.2°C per decade. The best estimates of long-term warming over the twenty-first century is between 1.5 and 4.5°C depending on the rate of economic growth and our greater or lesser dependence on fossil fuels.

Models also predict that climate change will affect the physical and chemical characteristics of the oceans. Estimates of the rise in sea level during the twenty-first century range from 30 to 40 cm (maybe 10–20 cm higher depending on the rate at which Greenland and Antarctic ice sheets melt), more than 60% of which will be caused by expansion of ocean water as it warms. Addition of vast quantities of fresh melt water will also change the nutrient content of the ocean as well as, possibly, its distribution by altered ocean currents.

With these facts and projections in mind, what can be said about how plants are adapting to global warming?

C_4 PLANTS

Earlier, it was noted that C_4 plants are largely unaffected by increase in CO_2 when compared to C_3 plants because they lack photorespiration. But C_4 plants are favored over C_3 plants in warmer climates. Will climate change favor C_4 plants by allowing them to grow more quickly?

In one experiment, photosynthesis was measured in groups of C_4 plants grown at moderate (25/20°C, day/night) or high (35/30°C, day/night) temperatures for an extended period of time. A major finding was that the long-term rates of photosynthesis in the higher temperature regime were not as high as predicted from earlier short-term laboratory measurements on isolated leaves taken from the same plants.

When examined in greater detail, the plants grown at the higher temperature were found to have lower than normal amounts of particular proteins in their chloroplasts, causing them to have a lower rate of photosynthesis. It was realized that the plants maintained a balance between the various photosynthetic components at each growth temperature by making selective changes to their chloroplast proteins, thus restricting the rate of photosynthesis, to balance allocation of stores of nitrogen.

Thus, as in the case of C_3 plants at higher CO_2 concentrations, when C_4 plants acclimate to higher temperature they also moderate increases in photosynthesis in favor of conserving nitrogen for their overall growth requirements.

The lesson learned

As in the case of CO_2, the lesson learned here is that information about growth is more relevant than that on photosynthesis when assessing how temperature affects plants.

FLOWERING AND TEMPERATURE

Evidence is mounting that global warming has changed the time at which first flowers appear by between 2 and 45 days in the past 30 years or so, depending on species and location. The average temperature during the month or two preceding flower opening seems to be of greatest importance.

One well documented case is that of over 500 plant species near Concord, Massachusetts, USA, beginning with observations between 1852 and 1858 by the naturalist, **Henry David Thoreau**, of *Walden* fame, and continuing through the efforts of others until 2006. Between 1852 and 2006, the Concord region warmed by 2.4°C owing to climate change and urbanization. Using the commonest 43 species recorded originally by Thoreau, plants in 2006 flowered 7 days earlier on average than in the middle of the nineteenth century.

In a more recent study, the average first flowering date of 385 British plant species advanced by 4.5 days between 1992 and 2002 compared with the previous four decades. Sixteen percent of these species flowered earlier than previously, with an average advancement of 15 days in the decade.

In general, it seems that annuals are more likely to advance their flowering than perennials, and that insect-pollinated species have more accelerated times of flowering (*phenologies*) than those pollinated by wind.

The consequences

These rapid changes may have serious consequences for the ecosystems in which they occur. For example, natural resource allocations between species may be altered as patterns of times of flowering change among them; if pollinators are involved, will they be available for pollination or be there in sufficient numbers if the time of flowering is altered significantly? Dislocations such as these could lead to changes in the size, species richness, and genetic diversity of plant communities.

Some plant species are incapable of coping with sudden climate change. In one survey of 80 tree species in the eastern United States, roughly half (36 of 80) were judged to have the potential to spread at least 100 km further north, including seven which might move more than 250 km. Another 30, it was predicted, would decrease their range by at least 10% in response to temperature change.

SUMMARY

- As the volume of scientific data on changes to the environment of the Earth grows, the greater is the confidence that human activities are the main cause of these changes
- The examples discussed here of the interactions between climate change and plants provide just a few snapshots of the importance of understanding how plants are reacting physiologically to global alterations in the biosphere. After all, responses of plants to factors such as CO_2, nitrogen, sulfur, ozone, water, and temperature, among many others, manifest themselves at the working, cell, level
- The future consequences of these webs of interaction and of interference by humans in the natural processes and cycles established over eons of time are difficult to predict
- What is already clear, however, is that the rapid changes already imposed on the biosphere by human activities are having profound effects on plants. If allowed to continue accelerating at present rates, they are certain to bring about further significant changes, the importance of which can only be inferred and deduced, not predicted with certainty.

Appendix   Genetic engineering: an essential tool in a rapidly changing world

Attempts in the past to create genetically engineered (GE) crops have led to a great deal of negative publicity about and considerable public antipathy towards the technology; this is unfortunate. Improvements to the technology have moved well beyond the first generation of GE crops, those with resistance to herbicides. More recent advances in this field have considerable relevance to several topics covered here in Parts IV and V; further advances now in the early stages of exploration and exploitation promise to transform the world as we know it.

One major driver in the future of GE agriculture will be climate change. The number one tool that all living organisms use to respond to alterations to their environment which occur naturally over time is genetic change, which is adaptive. But the kind of genetic change dependent on simple mutations of existing endogenous genes will not bring about adaptation fast enough to meet the challenges posed by rapid climate change in the case of crop species. GE is the only technology that is both sufficiently rapid and targeted to bring about adaptation to and mitigation of the human forcing of changes to global and regional environments in crops.

For example, nitrous oxide, a greenhouse gas some 300 times more harmful than CO_2, is percolating into the atmosphere from soil, to which we are adding increasing amounts of nitrogen fertilizer, worldwide. Varieties of crops now being genetically engineered for more efficient uptake of nitrogen promise to reduce the need for emendation of soil with fertilizer.

GE herbicide-tolerant crops, of which there are now several, are helpful in no-till agriculture. Farmers using these crops no longer need to till their land to control weeds before re-seeding. As a consequence, they bring about improvement to soil structure, greatly reduce erosion, and allow the soil to store more water. There are also major climate benefits from no-till practices: the soils of the Earth are estimated to hold about 1.7 Tt of carbon; this compares to 0.6 Tt held in living plants and 0.83 Tt in the atmosphere (see Chapter 15, Figure 15). Tilling releases into the atmosphere as CO_2 the carbon stored in soil. Cultivated soil may lose half of its organic carbon over time through tilling; no-till will not only maintain soil carbon but will result in restoration or net sequestration of it into depleted soils until they reach wildland levels.

The main ecological advantage of the GE crops, Bt corn and Bt cotton (Bt = *Bacillus thuringiensis*, a common soil bacterium which contains a unique toxin lethal to the larvae of the butterflies and moths that normally feed on crops such as corn and cotton), is that they reduce pesticide use significantly. The capability of producing Bt toxin engineered into these crops has proved highly effective in controlling pest infestations. This, in turn, not only allows a farmer to use less pesticide but also reduces the farmer's use of fossil fuel for applying chemicals to a crop, thus lowering greenhouse gas release.

Widespread advances are being made in genetically engineering into crops not just herbicide and pest resistance but also pathogen resistance, stress (drought, salinity, cold, heat) tolerance, and higher yield and quality, all of which take on added importance as climate change continues to advance.

But above and beyond all these initiatives in importance may be the growing interest in manipulating the microbes of the Earth. Microbes are not only the most ancient of life forms but also the most versatile. As we have seen in previous chapters, microbes play central roles in converting the key elements of life – carbon, oxygen, nitrogen, and sulfur – into forms that other organisms can access. They even far outdistance plants in contributing to the photosynthetic capacity of the planet, albeit mainly in aquatic, not terrestrial ecosystems. Their combined activities can be said to affect the chemistry of the entire globe through their ability to erode rocks, free metals from bound states, transform inorganic to organic forms, and break down even the toughest chemical compounds. They are ubiquitous and prolific.

It is interesting to note that the genes used for the first two commercial GE crop traits, glyphosate tolerance and Bt, both came from microbial sources, not from the genomes within the crops themselves. The vastly bigger genetic diversity among microorganisms than among crop genomes is likely to lead to a continued use of microbes as a source of genetic adaptation for crops.

Growing understanding of the mind-blowing versatility and capability of microbes has led to the advent of a new field, **synthetic biology**, the ultimate goal of which is "to design and build engineered biological systems that can process information, manipulate chemicals, fabricate new materials and structures, produce energy, provide food, and maintain human health and our environment" (Wikipedia). The potential of synthetic biology will likely, one day, make the GE agriculture of today appear pale and insignificant by comparison.

Our past use of genetic engineering may have been clumsy, but its future is enormously important and consequential. We will certainly need all the tools we can muster to respond to new and rapidly developing alterations to our global environment as they appear.

BIBLIOGRAPHY: PART V

Beerling, D. (2007). *The Emerald Planet: How Plants Changed Earth's History.* Oxford: Oxford University Press. *The key influence of plants on evolution of the planet over geological time.*

Beerling, D. J. and Berner, R. A. (2005). Feedbacks and the coevolution of plants and atmospheric CO2. *Proceedings of the National Academy of Sciences USA,* **102,** 1302–5. *Investigations into links in the evolution of terrestrial plants, CO_2, and climate over the Past 500 million years.*

Bekker, A., Holland, H. D., Wang, P.-L. *et al.* (2004). Dating the rise of atmospheric oxygen. *Nature,* **427,** 117–20. *When the level of oxygen began to increase in the Earth's atmosphere.*

Bernacchi, C. J., Kimball, B. A., Quarles, D. R., Long, S. P. and Ort, D. R. (2007). Decreases in stomatal conductance of soybean under open-air elevation of [CO2] are closely coupled with decreases in ecosystem evapotranspiration. *Plant Physiology*, **143**, 134–44. *How much evapotranspiration is changed in a particular crop when the atmosphere is enriched with CO_2.*

Berner, R. A. (1999). Atmospheric oxygen over Phanerozoic time. *Proceedings of the National Academy of Sciences USA*, **96**, 10955–7. *Significant variations have occurred in the planet's atmospheric [O_2] over geological time.*

Betts, R. A., Boucher, O., Collins, M. *et al.* (2007). Projected increase in continental runoff due to plant responses to increasing carbon dioxide. *Nature*, **448**, 1037–41. *The effect of increasing atmospheric [CO_2] on the hydrological cycle.*

Biello, D. (2010). Climate numerology. *Scientific American*, **302**, 14. *Scientists at the December, 2009 meeting on climate change try to determine what might be a "safe" level of atmospheric CO_2.*

Bradley, A. S. (2009). Expanding the limits of life. *Scientific American*, **301**, 62–7. *How and where life might have evolved deep in the ocean far away from the influence of sunlight.*

Brand, S. (2009). *Whole Earth Discipline: An Ecopragmatist Manifesto*. New York: Viking. *Provocative, thought-provoking, and scientifically rigorous views on the future of mankind in a rapidly changing world.*

Britannica Guide to Climate Change: An Unbiased Guide to the Key Issue of our Age. (2008). Philadelphia: Running Press. *The scientific evidence for climate change and possible solutions to it.*

Caldeira, K. and Wickett, M. E. (2003). Oceanography: anthropogenic carbon and ocean pH. *Nature*, **425**, 365. *Quantifies changes to ocean pH that could result from continued increases in atmospheric CO_2.*

Clark, S. and Ravilious, K. (2008). Unknown Earth: our planet's seven biggest mysteries. *New Scientist*, **199**, 28–35. *Answers questions about the origin and physical and chemical composition of the planet, and from where life might have come.*

Collins, W., Colman, R., Haywood, J., Manning, M. R. and Mote, P. (2007). The physical science behind climate change. *Scientific American*, **297**, 65–73. *These participants in the IPCC 2007 report summarize the arguments and discuss the uncertainties about climate change.*

Copley, J. (2001). The story of O. *Nature*, **410**, 862–4. *Considers the difficult question of why the early Earth appears to have lacked oxygen for so long.*

Cowie, J. (2007). *Climate Change: Biological and Human Aspects*. Cambridge: Cambridge University Press. *A textbook covering past, present, and possible future climate change from a biological, ecological, and human perspective.*

DeLucia, E. H., Hamilton, J. G., Naidu, S. L. *et al.* (1999). Net primary production of a forest ecosystem with experimental CO_2 enrichment. *Science*, **284**, 1177–9. *Controlled experiments demonstrating increases to primary productivity with increased CO_2 that are restrained by other factors in the long term.*

Duce, R. (2008). Texas A&M University (May 16). Atmosphere threatened by nitrogen pollutants entering ocean. *ScienceDaily*. http://www.sciencedaily.com (accessed July 11, 2008). *Increased removal from the atmosphere of CO_2 because of higher ocean productivity due to nitrogen pollution is partially offset by the release of nitrous oxide gas, which also forms in the ocean.*

Dwyer, S. A., Ghannoum, O., Nicotra, A., and Von Caemmerer, S. (2007). High temperature acclimation of C_4 photosynthesis is linked to changes in photosynthetic biochemistry. *Plant, Cell and Environment*, **30**, 53–66. *Reduced photosynthesis in experiments at higher temperatures was as a result of precise changes in chloroplast photosynthetic components.*

Field, C. B. (2001). Plant physiology of the "missing" carbon sink. *Plant Physiology*, **125**, 25–8. *Illustrates that limits to increases in growth of plants as atmospheric CO_2 levels rise may develop after several years.*

Fox, D. (2007). Saved by the trees? *New Scientist*, 27 October, 42–6. *Studies have shown that early increases in growth rate of trees in some tropical forests due to higher temperatures do not continue over the long term.*

Hardt, M.J. and Safina, C. (2010) Threatening ocean life from the inside out. *Scientific American*, **303**, 66–73. *Highlights the effects of increasing CO_2 emissions on ocean acidification, imperiling the growth and reproduction of plant and animal life.*

Hawkesford, M. J. and De Kok, L. J. (eds) (2007). *Sulfur in Plants: An Ecological Perspective.* Dordrecht: Springer. *Comprehensive treatment of sulfur and the biosphere.*

Hazen, R. M. (2010). Evolution of minerals. *Scientific American*, **302**, 58–65. *Follows the evolution of the formation of minerals on our planet and the influence of life on the process.*

Illinois State Water Survey: Nitrogen Cycles Project (2008). *Human Influences on the Nitrogen Cycle.* http://www.sws.uiuc.edu (accessed July 4, 2008). *A valuable source of information about the complexities of nitrogen release into the biosphere and nitrogen cycling.*

International Panel on Climate Change (IPCC) reports. http://www.ipcc.ch/ (accessed July 25, 2010). *Here can be found all four of the IPCC reports to date.*

Jacobson, M. C., Charlson, R. J., Rodhe, H. and Orians, G. H. (2000). *Earth System Science From Biogeochemical Cycles to Global Change.* San Diego: Academic Press. *A textbook which is a source of information about the biogeological cycles and climate change, among many topics.*

Kepler, F. and Röckmann, T. (2007). Methane, plants and climate change. *Scientific American*, **296**, 52–7. *Find that plants emit the greenhouse gas, methane, into the atmosphere through their stomata.*

LeBauer, D. S. and Treseder, K. K. (2008). Nitrogen limitation of net primary productivity in terrestrial ecosystems is globally distributed. *Ecology*, **89**, 371–9. *An analysis of the* *results of 126 experiments across the globe showing the strong interaction between the nitrogen and carbon cycles and how they affect one another.*

McAuliffe, K. (2008). Ocean reflux. *Discover*, July 28–37. *How much CO_2 is entering the oceans because of human activities and how this is affecting ocean pH and organisms.*

Miller-Rushing, A. J. and Primack, R. B. (2008). Global warming and flowering times in Thoreau's Concord: a community perspective. *Ecology*, **89**, 332–41. *Comparison of flowering times near Concord over the last ca. 150 years.*

Minorsky, P. V. (2002). The hot and the classic. *Plant Physiology*, **129**, 1421–2. *Advances in average first flowering dates among some British and American plants in the last half of the twentieth century and how these changes might affect interactions with pollinators and predators.*

Morison, J. I. L. and Morecroft, M. D. (eds) (2006). *Plant Growth and Climate Change*. Oxford: Blackwell. *Collection of articles on effects of changes to atmospheric CO_2, temperature, and water availability on plant life.*

Morison, J. I. L. (1987). Plant growth and CO_2 history. *Nature*, **327**, 560. *Morison comments on the effect of historically varying atmospheric CO_2 concentrations on plant growth and stomatal density, including the work of Woodward (see below).*

Raven, J. A. and Edwards, D. (2001). Roots: evolutionary origins and biogeochemical significance. *Journal of Experimental Botany*, **52**, 381–401. *Chronology of the first appearance and growth in importance over geological time of roots in biogeological cycling of, for example, soil minerals.*

Raymond, J., Zhazybayeva, O., Gorgarten, J. P., Gerdes, S. Y. and Blankenship, R. E. (2002). Whole-genome analysis of photosynthetic prokaryotes. *Science*, **298**, 1616–20. *That, in the bacteria examined, photosynthesis did not evolve via a linear path of incremental change but more rapidly by the swapping among them of blocks of genetic material known as horizontal gene transfer. These results have profound implications for the evolution of complexity in organisms more generally.*

Ricardo, A. and Szostak, J. W. (2009). Life on Earth. *Scientific American*, **301**, 54–61. *Some fresh clues about how life might have arisen de novo.*

Saleska, S. R., Didan, K., Huete, A. R. and da Rocha, H. R. (2007). Amazon forests green-up during 2005 drought. *Science*, **318**, 612. *Mapping the resilience of photosynthesis in tropical forests to drought using data from satellites.*

Schlesinger, W. H. (1991). *Biogeochemistry: An Analysis of Global Change*. San Diego: Academic Press. *The origin of chemical elements on Earth and evolution of geochemical and biogeochemical cycles.*

Schneider, D. (2006). That other greenhouse gas. *American Scientist*, **94**, 504–5. *The rise in atmospheric methane levels has ceased: implications for global warming.*

Sitch, S., Cox, P. M., Collins, W. J. and Huntingford, C. (2007). Indirect radiative forcing of climate change through ozone effects on the land-carbon sink. *Nature*, **448**, 791–4. *Effect of tropospheric ozone on plant stomata affects both the ability of the plant to take up CO_2 and its water balance.*

Smil, V. (2002). *The Earth's Biosphere*. Cambridge, MA: MIT Press. *Valuable source of information about the contributions of V. I. Vernadsky, among many other aspects of the biosphere.*

Smil, V. (2001). *Cycles of Life: Civilization and the Biosphere*. Scientific American Library. New York: Freeman. *Major source of information on G. T. Carruthers, V. I. Vernadsky, biogeochemical cycles, and the biosphere.*

Tausz, M., Grulke, N. E. and Wieser, G. (2007). Defense and avoidance of ozone under global change. *Environmental Pollution*, **147**, 525–31. *Cause and effect problems in determining how tropospheric ozone affects plant performance.*

The New Science of Metagenomics: Revealing the Secrets of our Microbial Planet. (2007). Report of committee. Washington: National Academies Press. *Provides details of the essential nature to all life of microbes and their future manipulation to our advantage. A free copy of this book can be downloaded at* http://books.nap.edu/

Townsend, A. R. and Howarth, R. W. (2010). Fixing the global nitrogen problem. *Scientific American*, **302**, 64–71. February. *Growing global use of nitrogen fertilizer is damaging the environment and threatening human health. Can we fix this problem?*

Tuba, Z. and Lichtenhaler, H. K. (2007). Long-term acclimation of plants to elevated CO_2 and its interaction with stresses. *Annals of the New York Academy of Sciences*, **1113**, 135–46. *Comprehensive review of how plants use their stress-coping mechanisms to deal with the negative aspects of elevated CO_2.*

Vernadsky, V. I. (2005). The renaissance of V. I. Vernadsky. *The Geochemical News – Newsletter of the Geochemical Society*, October, 2005. *A detailed account of Vernadsky's life and contributions.*

Visoly-Fisher, I., Daie, K., Terazono, Y. *et al.* (2006). Conductance of a biomolecular wire. *Proceedings of the National Academy of Sciences USA*, **103**, 8686–90. *How plants, growing under hot, dry, sunny conditions, protect themselves against excessive energy from the Sun.*

Woodward, F. I. (1987). Stomatal numbers are sensitive to increases in CO_2 from pre-industrial levels. *Nature*, **327**, 617–18. *A combination of correlative and causative studies to demonstrate that there is a close link between atmospheric CO_2 concentration and stomatal numbers.*

Wullschleger, S. D. and Strahl, M. (2010). Climate change: a controlled experiment. *Scientific American*, **302**, 78–83. *Scientists around the world are manipulating CO_2 levels and temperatures in a variety of ecosystems in an attempt to determine how such changes affect the biosphere.*

Epilogue

Green plants dominate our planet yet are often taken for granted. For many people, they are merely the passive aspect of a beautiful landscape, the "backdrop" against which animals exist. For others, they are essential to their lives but are there simply to be exploited for food, fodder, fuel, furniture, clothing, transport, recreation, health purposes, and protection without thought being given to their unique qualities as living things in their own right.

In the first four Parts of this edition, I have attempted to provide some insights into the very different world of green plants. Their lives are lived at a different pace from ours, which may be one reason why we so often forget that they are living organisms capable of doing so many of the things we also do. Like us, but in their own way, they can see, they can count, they can communicate with one another, they can be sensitive to the slightest touch and they can tell time with considerable precision. But they accomplish all of these things on a different timescale from most animals. Their very slowness deceives us into believing that they do not do much at all.

We should not forget, however, that green plants are unique among all organisms on Earth in that they alone have the means to use light as a source of energy. The very substance which renders them green, the pigment chlorophyll, puts them in the position of being the very foundation of our biosphere. They, by and large, sustain most of the remainder of life on this planet with the exception of the curious, fascinating organisms that derive the energy they need from deep in the Earth rather than from the Sun. Without green plants, our globe would be a much less vibrant place.

Inwardly, green plants are highly active, wonderfully complex chemical factories. The sugars they form with the aid of light energy in photosynthesis are just the first few of many thousands of chemicals that plants manufacture for their own use. After all, plants must grapple with many of the same problems as animals. First and foremost, they must feed and water themselves to sustain their growth. They must reproduce to ensure their survival into the next generation. To survive at all, they must combat their enemies which are numerous given that they are the primary source of food upon which all other organisms depend. They have to compete with their neighbors for space in which to live and gather nutrients for themselves. They must endure the elements to which they are exposed at all times given their relative immobility. The myriad chemicals they produce serve as essential agents in one or more of these functions.

Plants are, in some ways, more successful organisms than animals. They preceded the latter onto the land, they thrive in places where animals find it difficult or impossible to survive, and they can grow bigger and live longer than animals; and, in the final analysis, animals are totally dependent on them.

Today, increasingly, plants face additional problems created by human activities, such as global warming, pollution of land, water, and air, and ocean acidity. In Part V, some examples of the additional challenges that plants face and their responses to them are provided to demonstrate some of the ways in which plants are, or are not, capable of adapting to rapid change caused by human forcing of the environment. In the absence of sufficient information it is impossible to predict with confidence the limits of plant adaptation to change. Regardless, their central importance to the maintenance of all life on the planet is reason enough to spend time learning something of what makes them unique, why they are so successful, where they live and ... how they work!

Index